COMETS!

Join David J. Eicher in this fast-paced and entertaining journey through the history, present, and future of these important yet mysterious cosmic bodies. From ancient times, humans have been fascinated by "broom stars" and "blazing scimitars" lighting up the sky and moving against the fixed background of stars. The Great Comets of our time still receive in-depth attention – ISON, Hale-Bopp, Hyakutake, West, and others – while recent spacecraft encounters offer amazing insight into the earliest days of the solar system.

In this guide you will discover the cutting-edge science of what comets are, how they behave, where they reside, how groups of comets are related, and much more. The author carefully explores the ideas relating comets and life on Earth – and the danger posed by impacts. He finishes with practical how-to techniques, tips, and tricks of observing comets and even capturing your own images of them successfully.

David J. Eicher is editor in chief of *Astronomy* magazine, the world's largest publication on the subject. He is president of the Astronomy Foundation, the telescope industry's first-ever trade association. He is author of 17 books on science and history, and at age 15 he founded a magazine on observing galaxies, clusters, and nebulae, *Deep Sky Monthly*. An avid observer of astronomical objects for more than 35 years, he was honored in 1990 by the International Astronomical Union with the naming of minor planet 3617 Eicher.

COMETS!

Visitors from Deep Space

DAVID J. EICHER
Astronomy magazine

CAMBRIDGE
UNIVERSITY PRESS

32 Avenue of the Americas, New York NY 10013-2473, USA

Cambridge University Press is part of the University of Cambridge.

It furthers the University's mission by disseminating knowledge in the pursuit
of education, learning, and research at the highest international levels of excellence.

www.cambridge.org
Information on this title: www.cambridge.org/9781107622777

First published 2013

Printed in the United States of America

A catalog record for this publication is available from the British Library.

Library of Congress Cataloging in Publication data
Eicher, David J., 1961–
Comets! : visitors from deep space / by David J. Eicher.
pages cm
Includes bibliographical references and index.
ISBN 978-1-107-62277-7 (pbk.)
1. Comets. I. Title. II. Title: Visitors from deep space.
QB721.E355 2014
523.6–dc23 2013019975

ISBN 978-1-107-62277-7 Paperback

This work is dedicated to several astronomers I had the privilege of interacting with in my youth and who left a great impression on me. I will always be grateful.

Bart Bok
Brian Marsden
Carl Sagan
Carolyn Shoemaker
Gene Shoemaker
Clyde Tombaugh
Gérard de Vaucouleurs

Contents

Figures

Plates

Plates follow page xvi.

Foreword

With the appearance of two bright comets in the year 2013, sky watchers around the world are preparing to train their telescopes on this pair of wonders in the night. The starlike central portions surrounded by soft haze, and followed by gossamer tails, appear rarely enough in our lives that they surely deserve our full attention and our passion.

David J. Eicher's book brings the magical world of comets to life. It is not an arcane mathematical textbook but a celebration of these slowly wandering objects. It delves into Tycho's Comet of 1577, the comet that led to the great Danish astronomer's discovery that comets pass through the sky well beyond the orbit of the Moon. The book devotes considerable space to Halley, the most famous of comets and the first comet shown to be periodic.

Comets are famous not just for what they are, but also for what they can do. In 1994 Comet Shoemaker-Levy 9 performed the first major cosmic collision witnessed by humanity when it collided with Jupiter. My role in the discovery of that comet is a story that dates back to 1965, while I was a teenage camper at the Adirondack Science Camp near Lewis, New York. Our camp director asked that each participant display a project at the camp's annual science fair. He also insisted these projects need not be completed by the date of the fair; he wanted us to stretch our limits, to conceive and conduct a project that could last a lifetime, and one that could also fail. "Failure," he told us, "is the great teacher. If everything you do in science is a great success, then you probably haven't learned anything."

Although I completed a project that summer, I left camp dissatisfied. Two months later, while walking down to my high school French oral examinations, an idea suddenly hit me from out of the blue: I could begin a program of hunting for comets with a telescope. Inspired by the discovery of the Great Comet Ikeya-Seki of 1965, by the time I reached school I had already begun my plans. "Qu-est-ce que vous envisagez de faire une carrière?" (What do you want to do for a career?), asked Mr. Hutchinson, my examiner. Sitting up proudly in my chair, I replied, "Je veux découvrir une comète." (I wish to discover a comet.) Hutchinson sat up in his chair, stared at me, and asked in English, "How the hell do you expect to earn any money doing that?" We had a good laugh, and then he said, again in English, "Alright, but if you do not find a comet within the next 25 years, I will come back and lower your grade!" His timing was about right; I began my program on December 17, 1965, 62 years after the Wright brothers' flight, and discovered my first comet 19 years later, on November 13, 1984.

In 1989 I joined Gene and Carolyn Shoemaker's Palomar Asteroid and Comet Survey. Within the next few years we discovered 13 new comets, including 8 periodic comets, then numbered Shoemaker-Levy 1 through 8. On the night of March 23, 1993, I exposed two films that happened to contain the planet Jupiter. Two days later Carolyn, while exploring those films with her stereomicroscope, suddenly looked up. "I think I have found a squashed comet," she said. Gene looked at the films, then glared at me with a look of absolute puzzlement I had never seen before. When I got my turn, I noticed that each image showed two bars of fuzzy light with several tails pointing toward the top of the film, and each one surrounded on both sides by a thin trailed line.

Our discovery was just the beginning of the story. Over the next two years the comet figuratively whisked us off into space with it, and our fortunes were necessarily tied in with its headlong crash into Jupiter. During the week of the great impacts, July 16–23, 1994, we lived in Washington, D.C., and visited the Space Telescope Science Institute in nearby Baltimore each day. This comet taught us much about the splendid beauty of the night sky and the uncountable events featured within it.

Perhaps more importantly, as rare as this event was in our time, comet collisions were common in the early stages of the solar system. They may have accounted for much of the accumulation of water on Earth. In their many impacts with the early Earth, comets likely provided one additional thing. The materials they contain often sport particles of carbon, hydrogen, oxygen, and nitrogen. These CHON particles represent the simple alphabet of life. As CHON particles filter to the ground from direct impacts or close approaches, they can – over time – evolve into proteins, RNA, and DNA. When we look at a comet in the night, we may actually be looking at our own birthright.

These are the ideas that comets can engender. As you read this marvelous book, you may be learning about your own distant past.

David H. Levy

(David H. Levy is a Canadian astronomer, science writer, and comet hunter who is celebrated for his 22 comet and 41 asteroid discoveries. In 2010 he earned his Ph.D. from Hebrew University in Jerusalem.)

Preface

When I was 14, I fell in love with the universe. The discovery occurred with a one-time view of Saturn through a telescope at a local "star party." There was something so calm and comforting about gazing skyward at the twinkling dots spread across an inky black cosmos. Somewhere amid all the apparent serenity out in the universe things must be incredibly more complicated than they were for an earthbound kid. So in the cool spring of America's Bicentennial year, I found a new habit: taking my family's pair of 7×50 binoculars, grabbing a sleeping bag, and, with my dog in tow, wandering out from the edge of our neighborhood into a cornfield and lying down, taking long stares at the star fields and glows of the Milky Way above the southern Ohio landscape.

The sessions went from a few minutes at first, to hours after a couple of weeks. I might as well have been on the Moon; civilization was shielded, the neighborhood tucked away behind a low screen of trees, the sky fortunately dark, and my only companions Oscar the border collie, the occasional rustle of a squirrel or raccoon, and the deep beauty of the sky above. As Earth slowly rotated, I saw the universe's whole show – as far as we can see it from our place in the cosmos. Gradually, I learned constellations, recognized bright stars as my friends, and squinted toward the positions of fuzzy objects I couldn't quite make out – clusters of stars and glowing gas clouds in the Milky Way Galaxy – as seen through the old binoculars. Before I knew it, I had been taken into another world. I didn't know exactly why I'd gone, except that this world was alluring for the mystery and the grandeur of the vastness of space.

Soon I was heading out to the cornfield observatory on virtually every clear night. Warm clothes protected against the chill of a late-night dew; ultimately, an extension cord allowed a radio to carry soft music to the scene. Star atlases and the gentle glow of a red-filtered flashlight added ambience to what was becoming a junior science project. Each night I got to know my new friends better and better. There were Vega, Deneb, Altair, the North America Nebula, Coma Berenices. Lots of features presented themselves night after night. Once in a while a bright meteor streaked overhead. One particular meteor that spring was an absolute monster – a magnitude −4 fireball that lit up the ground in the early morning hours and, just as I was getting sleepy, jacked up my heart rate in an instant and perked up my dog.

Backyard astronomy became a way of life, an antidote to the uncertainty of the teenage years in high school. Before I knew it, I had joined an astronomy club, volunteered to write their observing column, and bought a telescope. It was an 8-inch Celestron Schmidt-Cassegrain scope, and after a year of observing with the binoculars, the scope added a whole new dimension to viewing distant objects. Now the fuzzy things scattered across the sky were resolved into sharp, impressive detail, as star clusters, nebulae, and galaxies far away from our own came into view. They too became well-known friends. There were so many of these so-called deep-sky objects, things beyond the solar system, that the astronomy club newsletter lacked room to write about them all. So by age 15 I began planning and publishing a little magazine called *Deep Sky Monthly*. It eventually attracted a following of about 1,000 subscribers and, before I knew it, I was publishing it throughout the rest of high school and into my college years.

Throughout my early years of observing the sky, I specialized in observing deep-sky objects, joining the staff of *Astronomy* magazine in 1982 and taking with me the retitled and quarterly magazine *Deep Sky*, which we published in Milwaukee for another decade. (Eventually it became clear that to move forward with *Astronomy*, the big title, I had to give up the smaller one.) But although my specialties were galaxies, planetary nebulae, and other fuzzy objects, I always kept an affinity for galaxy lookalikes – comets.

Comets very much resemble deep-sky objects as seen in a telescope's eyepiece. They move relative to the background stars, of course, whereas galaxies don't. But the visual similarities between comets and deep-sky objects drove the 18th-century French observer Charles Messier to produce the most famous catalog of sky objects in history, to separate the "nuisances" from the comets he cherished.

And comets loomed large in my early observing experiences. During my first big spring of observing the sky, Comet West (C/1975 V1) lit up the morning sky and became an instant classic, a brilliant spectacle. Soon after I arrived at *Astronomy*, the quirky comet IRAS-Araki-Alcock (C/1983 H1) appeared like a giant, luminous cotton ball as it quickly swept across the sky, passing close to Earth. I was fortunate enough to write most of the observing stories in *Astronomy* about Comet 1P/Halley, the most famous comet of all, during its 1985/6 apparition. And more recently, we all enjoyed big events like Comet Shoemaker-Levy 9 (D/1993 F2) slamming into Jupiter in 1994, and the Great Comets Hyakutake (C/1996 B2) and Hale-Bopp (C/1995 O1).

In late 2013, Comet ISON (C/2012 S1) is poised to be a Great Comet, putting on a brilliant show as it visits the inner solar system, pulled by the Sun. The universe has so much to offer and really changes lives with an understanding of the immensity and grandeur of all that awaits us "out there." I hope this book can play some small part in your discovery or appreciation of comets and the cosmos at large.

Acknowledgments

Many people's help goes into the making of any book, and such is the case with this work. It's particularly true when the timescale is so short: Spurred on by the discovery of Comet ISON, having waited for a bright comet for more than 15 years, I thought of doing a book on comets on the New Year 2013. After I received interest from the folks at Cambridge University Press, we all realized that we had just three months to put a book together, and the good folks at the Press wanted the better part of 100,000 words. This is my 17th book, and it has been far more of a scramble than any previous project. That I finished on time is due as much to many others who helped me as it is to hours of solitary writing.

First and foremost, I want to thank my wife, Lynda, and son, Chris, both great supporters and involved in their own worlds of discovery, Lynda as a teacher, helping to inspire young minds, and Chris as a chemistry student in college who will go on to help science move forward. And I thank my father, John, who at age 92 is still going strong and delights in discussing the philosophy of science at length once or twice a week. Thanks also to my sister, Nancy, for taking such good care of him.

I want to thank my friend David H. Levy, one of the great figures in the world of comets, for contributing his Foreword. I've known David since I was 17 years old, since the days he started contributing writings to my little publication *Deep Sky Monthly*. What a pleasure it has always been to know David.

Thanks also go in large part to Vince Higgs, my editor at Cambridge University Press, who has been a real pleasure to work with. I also want to thank Rachel Ewen, publicist at Cambridge, who worked tirelessly to get me involved in talking and writing for some others about comets. The production cycle on this book has been brief, but good fun every step of the way, thanks to Vince and Rachel.

I also want to thank the many photographers who generously contributed their sensational comet images to this project. Their spectacular success amazes me and is one reason why I've always stuck to visual observing. They are Guillermo Abramson, Luis Argerich, Paulo Candy, John Chumack, John A. Davis, Wayne England, Bill and

Sally Fletcher, Mike Holloway, Michael Jäger, Philip Jones, Brian Kimball, Martin Moline, Jack Newton, Damian Peach, Dennis Persyk, Gerald Rhemann, Dean Salman, Mike Salway, Chris Schur, Richard Simon, Michael Stecker, Jose Suro, Babak Tafreshi, Don Taylor, and Craig and Tammy Temple.

And a special thank you to Michael Bakich, *Astronomy* magazine's photo editor, for helping me to select some images.

At the tag end of the 1980s I wrote a book for Cambridge, a compilation of deep-sky observing stories called *The Universe from Your Backyard*. It's been 24 years since the publication of that book. I hope my next book for Cambridge will come a little sooner than that.

Plate 1. Comet Hale-Bopp (C/1995 O1) imaged from Austria on March 27, 1997, with a 190/255/435-mm Schmidt camera and Kodak Gold ISO 400 film, in an 8-minute exposure. Credit: Gerald Rhemann.

Plate 2. Comet Hale-Bopp appeared as a true spectacle in late March 1997, showing lovely blue ion and wide, fanned dust tails. Credit: Michael Stecker.

Plate 3. History's most famous comet, 1/P Halley, returns to the inner solar system every 76 years, as seen in this image made from Australia in 1987. Credit: Richard Wainscoat.

Plate 4. 17P/Holmes, a rather ordinary periodic comet, underwent an outburst in October 2007, temporarily brightening by a factor of a half million, its coma inflating to a diameter greater than that of the Sun. This image was made with a 105-mm refractor at f/5 at ISO 400, on November 1, 2007. Credit: NASA, ESA, and Alan Dyer.

Plate 5. Noted for its spectacularly long tail, Comet Hyakutake (C/1996 B2) put on an incredible show for observers in 1996, just as Comet Hale-Bopp was steadily brightening. This image was shot with a 130-mm f/6 refractor, ISO 800 film, and a 30-minute exposure in April 1996. Credit: Michael A. Stecker.

Plate 6. In late 1965, Comet Ikeya-Seki (C/1965 S1) swung into the inner solar system, very close to the Sun, and lit up so brightly that it was briefly visible in the daytime sky. Credit: Roger Lynds/NOAO/AURA/NSF.

Plate 7. 217P/LINEAR, a periodic comet discovered in 2001 by members of the Lincoln Near-Earth Asteroid Research (LINEAR) program in New Mexico, delighted observers in 2009 when it crossed Orion, passing between the Horsehead Nebula (left) and the Orion Nebula (right). This image was shot on September 28, 2009, with a 180-mm lens at f/4 at ISO 800 and stacked frames. Credit: John A. Davis.

Plate 8. Discovered in late 2011, Comet Lovejoy (C/2011 W3) is a sungrazer that produced a long, ghostly tail, well positioned for Southern Hemisphere observers. This image was made on December 27, 2011, from South Australia. Credit: Wayne England.

Plate 9. Comet Lulin (C/2007 N3), discovered in China in 2007, developed a strong greenish color due to high levels of diatomic carbon in its coma, and a stunning antitail (left) as well as a disconnected, withered dust tail (right). The combination made it one of the most usual comets ever observed. This image was made on February 22, 2009, with a 106-mm refractor, CCD camera, and 48 minutes of exposure time. Credit: Philip L. Jones.

Plate 10. Captured on January 21, 2007, Comet McNaught (C/2006 P1) shows an arcing tail spanning 45° in this image taken near Santiago, Chile. The image was made with a 45-mm lens at f/2.8, ISO 1600, and a 30-second exposure. Credit: Richard Simon.

Plate 11. Comet NEAT (C/2001 Q4) glows brightly, showing its coma and inner tail, in an image taken with the 0.9-m WIYN Telescope at Kitt Peak National Observatory near Tucson, Arizona, on May 7, 2004. Credit: NASA, NOAO, NSF, T. Rector (University of Alaska-Anchorage), and Z. Levy and L. Frattare (STScI).

Plate 12. Comet PANSTARRS (C/2011 L4) showed a broad, yellowish dust tail on March 19, 2013, when it was captured with a 101-mm refractor, a CCD camera, and stacked exposures. Credit: Gerald Rhemann.

Plate 13. On March 12, 2013, Comet PANSTARRS (C/2011 L4) peeks through clouds over the William Herschel Telescope at La Palma in the Canary Islands, along with the young Moon. Credit: Babak Tafreshi, The World at Night.

Plate 14. This image made with the Hubble Space Telescope on July 18, 1994, shows the impact scars on Jupiter from fragments D and G of Comet Shoemaker-Levy 9 (D/1993 F2). The prominent G impact site shows a central dark spot 2,500 km across, surrounded by a thin dark ring spanning 7,500 km. Credit: H. Hammel. MIT, and NASA.

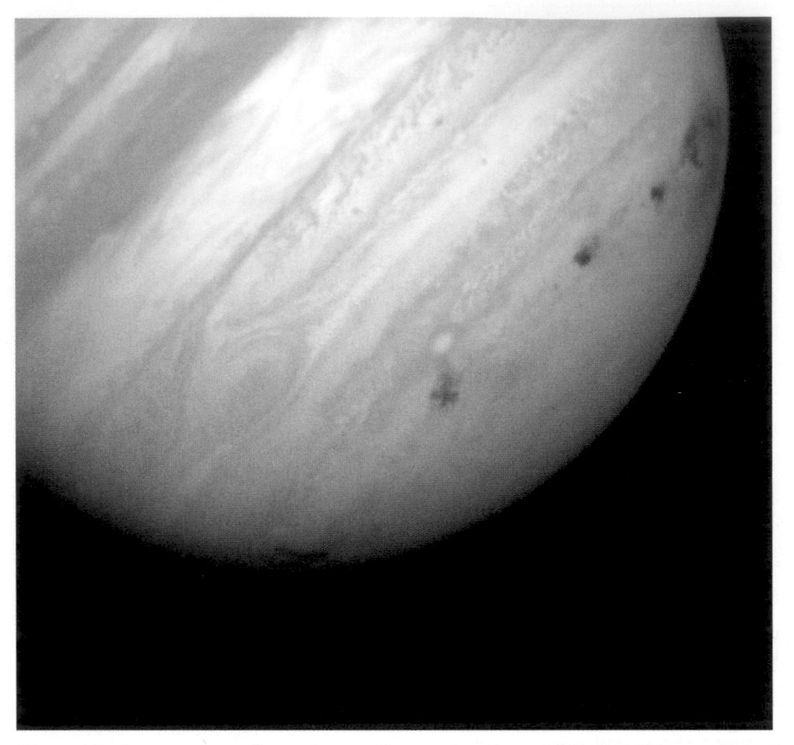

Plate 15. Impact scars from Comet Shoemaker-Levy 9 (D/1993 F2) are aligned across the cloud tops of Jupiter in this image made with the Hubble Space Telescope on July 22, 1994. Eight impact scars are visible, from left to right, the E/F complex, the star-shaped H site, N, Q1, small Q2, R, and the D/G complex on the right-hand limb. Credit: Hubble Space Telescope Comet Team and NASA.

Plate 16. The star comet of the 1970s, Comet West (C/1975 V1) blazes brightly in the morning sky in the spring of 1976. Credit: ESO.

1

Strange Lights in the Sky

When I was young I fancied becoming a doctor. The allure of medicine, of diagnosing diseases, of understanding the complexity of the human body – it all seemed endlessly fascinating. It offered a universe of ideas and challenges you could lose yourself in that could help person after person through challenges with illness and health. And then, in the midst of that momentum, when I was 14, in my little southwestern Ohio town, I went to a so-called star party.

Someone had set up a Criterion Dynascope 6-inch reflector, one of those telescopes with a long white tube and with the eyepiece fixed high at the upper end, and I peered in to take my first telescopic look at Saturn. That moment changed my life. Seeing the radiant light from Saturn's bright orange globe, encircled by golden orange rings, incised by a black gap, made me gasp. The pinpoint of a little saturnian moon hovered nearby. Everything just stopped. I was transfixed by the vision of another world – live, in real time – right before my eyes.

It was early 1976, and the crisp winter air was not yet ready to surrender to spring. Infected with this new awareness of the universe around me, I needed to find out everything I could – to take many more looks through telescopes, or in my case, my dad's pair of binoculars. Just as I was scrambling my first set of primitive equipment together, a friend called and gave me some promising news. "If you're getting into astronomy, you're in luck," he blurted out. "There's a bright comet that's gonna be amazing soon, but you'll have to get up early in the morning to see it!"

Along came another magic moment. Wandering out into the backyard, stepping across into the adjoining cornfield, and gazing up at the stars of Aquarius, I was thunderstruck at the sight. The icy cold air, dead silence of the early morning, and strange adventure of being out alone in a field before dawn added to the eerie, almost mystical sight that hovered over the planet. There, starkly visible in plain sight, like a shimmering sword hanging over Earth, was the bright glow of a comet with a fuzzy, starlike head and a long tail skirting upward and to the left. This was my first look at Comet West, the first look of many.

To someone who lived his whole life to that point on a "2-D planet," like most of us beset by issues of daily life, this was a dose of sudden magic. Who knew that you could simply walk out and so easily see such a range of incredible sights in the universe, far away from Earth? And not only was Comet West a spectacular sight, bright enough to be stunning in its odd and unexpected appearance, but it showed me in just a day or two that objects in the heavens changed rapidly. The comet altered appearance when viewed through a telescope and changed position in the sky from night to night. I was catching on that there's a whole lot more to this universe than I might have believed just a few days earlier.

Each day the comet rose in the early morning sky in the east, the tail peeking above the horizon first and then finally the head clearing the trees and moving up to complete the stunning portrait. Each morning it was fully visible in a dark sky before the creeping glow of dawn finally moved in and broke up the show. Here was a daily adventure, one that revealed the universe around us as a dynamic and unpredictable place. It demonstrated loudly that we inhabit just one little tiny spot in the cosmos, indeed even a small corner of our solar system. That late winter and early spring, Comet West became one of the Great Comets of the 20th century, peaking at magnitude –3, making it brighter than the planet Jupiter.

Strangely, you don't have to go very far back into history to reach the point when great thinkers believed comets were local phenomena, emissions of gas or smoke hovering in Earth's atmosphere. But then astronomers realized that if they were close, comets would be seen against slightly different star backgrounds from different places on Earth, and that didn't happen. The realization came on that comets are distant objects – at least much more distant than the Moon – moving through the solar system in very strange ways compared to the regular orbits of the planets.

So what exactly are comets, anyway? The answer requires looking at the way planetary scientists believe the solar system formed, some 4.6 billion years ago. The solar system consists of our Sun, a medium-sized star, and its attendant planets and other assorted debris, the whole collection being one of perhaps 400 billion stars and attendant small bodies in the Milky Way Galaxy. (And the universe contains some 125 billion galaxies astronomers know of – it's a rather large place.)

Scientists believe the solar system formed as a giant disk as gravity pulled material inward, eventually assembling enough mass to enable the Sun to "turn on" its nuclear fusion and begin life as a star. The so-called solar nebula, the disk formed and spun in rotation by gravity as the solar system coalesced, contained lots of material that didn't make it into the Sun itself. Some of this material was eventually driven off by radiation pressure from the Sun's intense energy, but some continued to join together by gravity, sticking little bits into larger bits, and building planets.

But Sun and planets alone do not make a solar system. Several distinct zones make up our star's system. The innermost zone contains the terrestrial planets, including Earth. Next comes the asteroid belt, a region of rocky debris containing thousands of subplanet-sized bodies that together make up less mass than Earth's Moon. Next come the giant planets, including Jupiter and Saturn. And the outer zone contains the comets, icy bodies of frozen gases and dust.

Most comets are far, far away and exist in several groups. Some comets are locked up with other debris in the so-called Kuiper Belt, a disk of icy bodies that extends from about 4.5 billion km out to 7.5 billion km from the Sun – in the region of Pluto and other dwarf planets. (The Sun itself spans a mere 1.4 million km.) Other, more remote comets exist in a huge shell surrounding the solar system called the Oort Cloud. Planetary scientists believe some 2 trillion comets may exist in the Oort Cloud, nearly all of which never make their way in toward the Sun (and into our skies). The Oort Cloud extends a staggering distance into deep space, perhaps as many as 1.5 light years from the Sun; that's 40 percent of the way to the nearest star beyond our Sun. The comets themselves are typically just a few kilometers across. And some other zones and families of comets exist too.

In later chapters, we'll explore the great complexity of comets, their origins, and where they live in detail, and you'll see how incredibly rare a thing it is for a comet to move into the inner solar system and become terrifically bright. We'll discover that asteroids and comets, not long ago believed to be two separate things, are now blurring the lines of their relationships. We'll absorb the findings from spacecraft missions and ground-based telescopes that have studied both bright and faint comets and expanded our knowledge of these mysterious visitors. We'll revisit how comets have affected human culture, how people have celebrated or dreaded them throughout history. And we'll examine the best ways to observe and photograph comets from your own backyard, or whatever dark-sky sites you prefer.

The appearance of a bright comet in Earth's skies is one of the most exciting astronomical events of all. In fact, no other type of astronomy-related happening comes close in getting new people interested in the night sky. Whenever a really bright comet appears, chatter rises, club memberships increase, attendance at star parties zooms, circulations of astronomy magazines climb, and new blood enters the hobby of astronomy.

It's happened that way time after time since astronomy became an organized hobby, most recently with the Great Comets Ikeya-Seki (1965), West (1975/6), Halley (1985/6), Hyakutake (1996), and Hale-Bopp (1996/7). Now more than 15 years have passed since the last terrifically bright, well-placed comet has graced the skies of Northern Hemisphere viewers. But the time may have come.

My friend David Levy, one of history's most successful comet hunters, has a favorite saying. "Comets are like cats," he claims. "They have tails, and they do precisely what they want." This underscores one of the great challenges with comets – their lack of predictability. The latest go-around occurred in September 2012, when astronomers discovered a potentially bright comet that could dazzle observers the world over in the fall of 2013.

On September 21, 2012, astronomers Vitali Nevski from Vitebsk, Belarus, and Artyom Novichonok of Kondopoga, Russia, captured images of a new fuzzy object in the sky. Their instrument of choice was the 16-inch Santel reflector at Kislovodsk Observatory in Russia, along with a program of automated asteroid detection called CoLiTec. The telescope is one of 18 dedicated by the Russian Academy of Science to detection and tracking of faint objects in the sky, the network collectively termed the International Scientific Optical Network (ISON).

When the Russian astronomers alerted others that they suspected a comet, astronomers at the Mount Lemmon Survey in Tucson, part of the Catalina Sky Survey, and astronomers at the Pan-STARRS telescope in Hawaii checked earlier images and also found the object. The next night, more observations were made by Italian astronomers at the Remanzacco Observatory, using another network, this one called iTelescope. The Minor Planet Center in Cambridge, Massachusetts, clearinghouse for such astronomical discoveries, announced the new comet on September 24, 2012, three days after its discovery at a terrifically faint magnitude of 18.8.

As with all comets, following its discovery and verification by other astronomers, the new fuzzy object received a designation, C/2012 S1, and its popular name would not be the name of one of the discoverers but, following international agreements, the search network abbreviation. So C/2012 S1 (ISON), or informally, Comet ISON (Figure 1.1), was born. (Many such search facilities have uncovered multiple comets, however, so care needs to be used in throwing around the term Comet ISON or those of other networks or surveys.)

ISON is exciting to astronomers because of its great potential as a so-called sungrazer – a comet that will swoop in very close to the Sun and therefore brighten dramatically. At perihelion, its closest approach to the Sun, the comet will pass a mere 1.8 million km from our star's glowing "surface." When this happens, on November 28/29, 2013, the comet could be dramatically bright, a significant fraction as bright as the Full Moon. But that will take place in a daytime sky, when the comet is only 1.3° northeast of the Sun.

Fortunately, the comet should be dazzling in a nighttime sky as well – to be more precise, in the early morning sky in mid-November. The comet could then shine as bright as the planet Venus and may well become the brightest comet ever seen by anyone now alive.

Another reason ISON's potential is exhilarating is that its orbit resembles that of another famous comet, C/1680 V1 (Kirch), which came to be called the Great Comet of 1680. Because the orbits are so similar, some astronomers have speculated they

Figure 1.1. The tiny fuzzball at the center of this image, shot on January 16, 2013, is 16th-magnitude Comet ISON (C/2012 S1), which observers hope will brighten dramatically by late in 2013. The imager used a 20-inch Ritchey-Chrétien scope, a CCD camera, and stacked exposures. Credit: Dean Salman/NOAO/AURA/NSF.

may have originated from the same parent body. If this is so, ISON may present a historic show as well. The Great Comet of 1680 was one of the brightest comets of the 17th century and was plainly visible during the day. And its distance from Earth at closest approach was nearly the same as ISON's will be. ISON will reach perigee, its closest passage of Earth, on December 26, 2013, some 63 million km from Earth.

At the turn of the New Year 2013, ISON glowed faintly at 16th magnitude as it floated among the stars of Gemini, near the bright twins Castor and Pollux. In addition to professionals at research observatories, amateur astronomers began to image the comet with a great sense of anticipation. Because of the orbital geometry of its path through the inner solar system, ISON will commence a big, semicircular, clockwise loop through the sky beginning in spring 2013, traversing Leo, Virgo, Scorpius, Hercules, and Ursa Minor by January 2014.

But the comet's great brightness will be a long time coming. By midsummer 2013 ISON will brighten to be an intriguing telescopic fuzzball; it likely won't be until early fall that ISON hits the range of being an impressive comet as viewed with binoculars. In October, the excitement will build as ISON's magnitude rises above 10, and sometime close to Halloween, the comet will become a naked-eye object.

Comet fever should grip the astronomy world – and maybe pop culture too – when ISON slinks across southern Leo and into Virgo during the first week of November.

By then, the comet will rise to 6th magnitude, and a few days later it will gain another magnitude and be visible with the eye alone from a suburban site. The comet then should increase by a magnitude every few days and will dazzle viewers who rise to see it in the early morning hours, perhaps 4 A.M. on into dawn. (That's the time slot occupied by Comet West during those memorable first few weeks of 1976.)

If predictions pan out, by about November 25 the comet will have become impressively bright, shining at negative magnitudes, and situated in eastern Virgo, approaching the border with Scorpius. November 28 is the comet's perihelion, its closest point to the Sun. ISON may then be as bright as Venus, or as much as 100 times brighter yet. If so, it will outshine everything in the sky save for the Sun and the Moon.

But remember that we're talking about the comet's total magnitude, its brightness if all of the light were compressed into a pointlike source. Because the comet is spread out over a large area, little areas of it will not appear as bright as Venus. But we're still talking about a comet that could cast shadows – a remarkable event that's unprecedented in our lifetimes.

At its brightest moment, ISON could shine at magnitude –9.5. By then it will be a daytime object a mere 1.3° from the Sun. This will make seeing the comet at its brightest difficult; trained observers who block out the disk of the Sun will be able to see it, but it will not be an easy observation when the comet is so close to the solar disk.

Comet ISON lies right in the head of Scorpius at perihelion and thereafter swings north toward Hercules. The first week of December should see it as a 1st-magnitude object with a sweeping tail, and Northern Hemisphere viewers will be well placed to see the comet as it slowly fades toward month's end. By January 8, 2014, the comet will be a mere 2° from Polaris, the North Star, and will have dimmed to about 6th magnitude, reaching the naked eye limit once again.

Of course predicting comet magnitudes makes for a dangerous game. The comet's orbit is well known, but assumptions about the comet's composition, how solid it is, its reflectivity, and how volatile its gases and dust are make the brightness of ISON uncertain. Recent astronomical history knows one great story of a comet that everyone believed would certainly be dramatically bright, and in the end it fizzled. That is the story of Comet C/1973 E1, Kohoutek.

Shortly after its discovery, Comet Kohoutek was touted as the "comet of the century." Among the prognosticators who believed Kohoutek would put on a spectacular show was Carl Sagan, not yet world famous as the creator of *Cosmos*, the book and television miniseries, but famous enough as a compelling scientist (and astronomy professor at Cornell) to appear on the *Tonight Show* alongside Johnny Carson.

Sagan predicted a sensational view of the comet as Kohoutek brightened late in 1973 and early in 1974. The comet, after all, promised a great deal to astronomers as they studied its orbit. It had been discovered on March 7, 1973, by Czech astronomer Luboš Kohoutek (1935–), who along with other astronomers found that the comet was a long-period object with a hyperbolic orbit that would carry it extremely close to the Sun. The date of perihelion was fixed as December 28, which would provide the world with an end-of-year, holiday spectacle.

Astronomers excitedly found the comet would pass close to Earth and quite close to the Sun, a mere 21 million km. The blowing of the horn about how fantastic Kohoutek would be ramped up expectations and created quite a flurry of attention in the popular media and culture, aside from Sagan's regular pronouncements.

The comet's effects ranged from the ridiculous to the sublime. David Berg, founder of the Children of God, predicted a doomsday event for January 1974. In December 1973, jazz musician Sun Ra put on a Comet Kohoutek show. The comic strip *Peanuts* featured the comet over a week-long span as Snoopy and Woodstock hid under a blanket from the mysterious light from the sky. The comet influenced musical works at the time or later by Pink Floyd, R.E.M., Journey, Kraftwerk, and Weather Report.

And the reasons for optimism were valid. The feeling was that, with such an orbit, Comet Kohoutek must be an Oort Cloud object, originating from far out in the solar system and therefore fresh, rich in volatile gas and dust that would stream off the comet as it warmed in the glow of sunlight like water vapor taking to the air on a foggy London morning. Astronomers believed the comet had never been to the inner solar system before and therefore was a good, solid object.

But as Kohoutek approached the inner solar system, it lagged significantly behind the predicted magnitudes. Comet Kohoutek seemingly fooled the experts on two counts: In hindsight, it may well have originated from the closer Kuiper Belt, not the distant Oort Cloud, and therefore could have had a relatively rocky composition with minimal volatile ices, gas, and dust. Moreover, rather than reflecting sunlight efficiently and developing significant tails spread across the sky, the comet partially disintegrated as it approached perihelion, prior to its closest approach to Earth. Thus, although it became a naked-eye comet, to many, Kohoutek was an outright dud.

The lesson is simple: No one can accurately predict a comet's brightness beforehand, even knowing its orbit well, because of many small but potentially important unknown factors. ISON will be a great sight: no doubt. Only by early 2014 will we all know whether it was really the comet of our lifetimes, the century, several centuries, or just another pretty good comet. But the upside with this discovery is that it could "fizzle" compared to the predicted magnitudes and still be a remarkably, perhaps historically bright comet. That's pretty encouraging. I urge you to follow the comet's progress in *Astronomy* magazine and on the magazine's Web site, www.Astronomy.com.

Well, whether it be the excitement over observing a bright comet or the anticipation of what might be to come, one question soon comes to mind: Just what exactly is a comet, anyway? The Greeks originated the word *kometes*, which translates to "long-haired," referring to what early observers thought of as "hairy stars" because of their observed glowing tails. Although most people think of comets as a "streak" of light in the sky, a comet is really a tiny body floating along some kind of orbit in the solar system. Comets are clumps of frozen ices, gas, and dirty rock the diameter of a small town – they span an average of 5 kilometers or so across – and only when this frozen chunk approaches the inner solar system and heats up from the Sun's warmth does it begin to outgas and produce a tail, becoming a spectacle in the sky.

Later chapters will describe the nature and physics of comets in detail, but for now, suffice it to say that its physical body is called the nucleus. The observational parts – coma, a hazy cloud of light surrounding the nucleus, and tails – arise from the solar heating as the comet approaches the Sun. The nucleus is really the physical being of the comet, and planetary scientists believe cometary nuclei formed in the proto solar system some 4.6 billion years ago as icy outlying material that did not gravitationally clump together into larger objects.

As solar system bodies go, comets are tiny. You could stack 2,500 of them side by side across Earth's equator. Yet they are plentiful – the deep recesses of the outer solar system may hold as many as 2 trillion comets. The bulk of a comet is frozen ices and gases; it is chiefly composed of water, carbon monoxide, carbon dioxide, formaldehyde, and methanol. Along with the frozen ices and gases are variable amounts of dust.

The unpredictable nature of comets as they approach the Sun and warm up results from their variable composition, the amounts of various gases and dust, the "freshness" of the comet – some have taken previous trips to the inner solar system – and the dynamics of the orbit. When the frozen block of comet approaches the Sun and warms, its ices begin to sublimate, transforming directly from a solid to a gas, and this creates the diffuse coma, which is surrounded by a large halo of hydrogen. The coma also contains dust grains liberated from being locked in the ice.

As the comet continues to warm, more gases escape and dust grains leap forth, and as the volume increases, the solar wind and radiation pressure from the Sun push these particles into a gas or plasma tail (typically bluish) and often a separate dust tail (white to yellowish in color), pointing away from the Sun.

Comets are one major type of small body in the solar system. The other consists of rocky bodies without frozen gases that mostly live in separate places – the asteroids. Together, comets and asteroids make up the vast majority of the small bodies in the solar system. For hundreds of years, astronomers classified comets and asteroids as two completely different creatures – apples and oranges. As we'll see later, however, at least in some cases, this distinction is becoming blurred as new discoveries are made.

One important feature that distinguishes comets is their peculiar orbital track around the Sun. They typically have large orbital eccentricities – orbits that differ markedly from circles – and high orbital inclinations, often tipping them at a strange angle compared with the orbits of the planets. Eccentricities are sometimes elliptical, sometimes parabolic, and sometimes even hyperbolic in the cases of comets that have been influenced by the giant planets and slung like pinballs in a crazy game of orbits. The angle of cometary orbits relative to the plane of the solar system is essentially without limits. Clearly, chaos in the early solar system was instrumental in setting up the paths of these celestial wanderers. Many are well behaved; others drop down at crazy angles like dive bombers; some even have so-called retrograde orbits, moving around the Sun in the opposite direction of Earth and the other planets.

As with all sciences, astronomy was for hundreds of years chiefly a game of classification. To help understand the orbits of comets – and where they might be coming from – planetary scientists created a dividing line for comets based on their orbital periods.

Long-period comets, those with periods of more than 200 years, are governed by highly elliptical orbits, and they reside at the outer limits of the solar system. These creatures spend their lives in a celestial deep freeze, a measurable fraction of the distance to the nearest star away, and move in only rarely and briefly to our part of the cosmos. In fact, an enormous shell of comets surrounds the solar system, and the distinguished Dutch astronomer Jan H. Oort (1900–1992) proposed in 1950 the existence of the source of these long-period comets, and the great sphere of cometary nuclei took on his name, the Oort Cloud.

An interesting class of long-period comets exists in the Kreutz Sungrazers. They are named for German astronomer Heinrich Kreutz (1854–1907), who demonstrated their relationships. This family of comets is characterized by orbits that carry the celestial visitors very close to the Sun, which sometimes makes them exceptionally bright. Sungrazers sometimes plow straight into the Sun or break apart as a result of the Sun's gravitational influence.

By contrast, short-period comets have orbits of 200 years or less and are further divided into two distinct groups. These objects are much closer residents of the solar system. They comprise the Halley-type comets, which have periods of 20 to 200 years, and the Jupiter-family comets, with periods of less than 20 years. The Halley class feature orbits that are randomized, just as the long-period comets do, but the orbits of the Jupiter-family comets are inclined more closely to the ecliptic plane, as are those of the planets.

Of the 2 trillion comets that may exist in the Oort Cloud, astronomers have observed and cataloged about 4,200. Some 1,500 of these are Kreutz Sungrazers and

484 are short-period comets. In recent years astronomers have even detected comets in extrasolar planetary systems, the first in observations of the Beta Pictoris system in 1987. Of the 10 so-called exocomets discovered to date, all were detected around young stars, and they may help tighten the picture of how solar systems form.

This is a pretty strong indicator of how fast astronomy is moving as a science. Several hundred years ago many of the planet's best thinkers believed that comets were atmospheric phenomena. Now we're observing comets that are dozens of light-years away.

Come to think of it, that's one of the things that struck me as a teenager, lying out in that field, gazing up at Comet West. Suddenly, after I learned a little about what comets are, it hit me. They hammer home the immensity of the cosmos. Yes, they are relatively nearby. But seeing them move from night to night – changing their place against the backdrop of the stars glistening behind them – is extremely powerful. I think it triggers something deep within the soul.

And that seems always to have been the case. The earliest records of cometary observations are from China and date from about the year 1000 B.C. Similar observations may have been made by inhabitants of the marshy land in southeastern Mesopotamia known as Chaldea. By about 550 B.C., Greek philosophers recorded comets as wandering planets. In his scheme of spherical shells making up the cosmos, Aristotle (384–322 B.C.) wrote in *Meteorology* (*ca.* 330 B.C.) that comets are residents of the lowest such sphere and called them "dry and warm" atmospheric exhalations.

Not only were comets viewed as local phenomena, but for centuries they were also taken as portents of doom, omens of some impending event, usually a disaster. Only with the writings of Thomas Aquinas (1225–1274) and Roger Bacon (*ca.* 1214–1294) did the notion that comets may not be lurking in Earth's atmosphere begin to step forward. But further intellectual work on the subject would really have to wait until the world emerged from the gloomy deep freeze of the Middle Ages.

Real progress on understanding comets stepped up when Paolo dal Pozzo Toscanelli (1397–1482), an Italian mathematician and astronomer, observed what would come to be known as Halley's Comet in 1456, along with a number of other comets during the previous and following decades. Observations made by Toscanelli and later by Danish nobleman and astronomer Tycho Brahe (1546–1601) began to define comets more precisely. Tycho's observations of Comet C/1577 V1 in particular demonstrated the comet's distance as being much farther away than the Moon.

In the late 17th century, German amateur astronomer Georg S. Dörffel (1643–1688) observed two bright comets in 1680 and 1681 and realized the two comets were one comet – C/1680 V1 – seen before and after perihelion, and that the comet had a parabolic orbit about the Sun. This provided ammunition for the great physicist Isaac Newton (1642–1726), who used his newfound theory of gravitation to demonstrate the comet moved in a giant elliptical orbit and passed only 230,000 km above the Sun's surface.

And then the astronomer whose name would almost become synonymous with comets entered the stage. English astronomer Edmond Halley (1656–1742) calculated the orbits of a dozen well-known comets and found that one in particular, the Great Comet of 1682 (1P/1862 Q1), was periodic. Further, he postulated the comet's period as 76 years (its period has varied between 76 and 79 years).

When in 1758 this comet was recovered as predicted by Halley, this time by German astronomer Johann G. Palitzsch (1723–1788), the first chapter of the story of cometary orbits fell together, neatly bundled. Newton's theory of gravitation was validated to a distance way out beyond the planet Saturn, and Halley's prediction rang true, solidifying the most celebrated name in comets, "Halley's Comet."

Astronomers continued to observe comets and refine their techniques for determining orbits over the 18th and early 19th centuries. To their puzzlement, scientists increasingly found that some comets had orbits like nice parabolas while others were contained in the inner solar system, closer to Earth than Jupiter. Astronomers believed that somehow the giant planet was gravitationally dominating the orbits of these comets, or perhaps even that Jupiter itself was ejecting the comets.

It was the French astronomer Pierre-Simon Laplace (1749–1827) who realized that a gravitational mechanism enabled Jupiter to capture these comets and influence their orbits, concentrating some of the short-period comets in one area. Thus, the Jupiter family of comets was born.

As telescopes improved, so did the opportunities for observing comets. The 1835 appearance of Halley's Comet afforded multiple observers the opportunity to make detailed drawings of the comet's structures, such as jets, streamers, and brighter and darker portions of the coma. German mathematician and astronomer Friedrich Bessel (1784–1846) observed the comet and supposed he saw particles streaming out of the comet's nucleus and being forced back into a tail that aimed away from the Sun. By this time, comets had become real – they were interacting with the solar system around them.

As the 19th century edged onward, astronomers discovered more bits of evidence that comets were important parts of the cosmos, not just meaningless debris. In 1866 and the year that followed, Italian astronomer Giovanni Schiaparelli (1835–1910) identified two annual meteor showers, the Perseids and the Leonids, with two comets – 109P/Swift-Tuttle and 55P/Tempel-Tuttle, respectively. The fact that these comets were losing particles that later intersected Earth's orbit, shooting into our atmosphere and creating glowing streaks, linked comets with meteors.

Around this same time, astrophysics entered the study of comets when Italian astronomer Giovanni Donati (1826–1873) and English astronomer William Huggins (1824–1910) made the first spectroscopic observations of comets. They found the spectroscopic bands seen in several comets and in a gas flame were similar, and they detected a broad continuum indicating that the comets were reflecting sunlight.

The two great early tools in the arsenal of astrophysics, photography and spectroscopy, soon became the standard for comet research. The English photographer William Usherwood (1821–1915) took the first photograph of a comet when he recorded C/1858 L1 Donati, in 1858. Good spectra of cometary tails were made soon after the turn of the 20th century, and a big event in 1910, the next apparition of Halley's Comet, gave astronomers the opportunity to produce some of the earliest papers on the physics of comets.

Leaps in understanding comets would have to wait. Not until the 1950s did several events take place that would push the understanding of comets forward. In 1950, American astronomer Fred Whipple (1906–2004) proposed the icy conglomerate model of a comet's structure and composition (now universally and lovingly known as the "dirty snowball" model). The seeds of this idea went back to the late 1930s, but Whipple was the first to put them all together. Astronomers hadn't yet understood how molecules were locked away in a comet's frozen nucleus. Arguments over the source of the gas and dust from a comet had originated nearly a century earlier, and astronomers still couldn't explain how a comet's nucleus, when warmed, could produce such a vast coma. Moreover, they couldn't understand how comets could be repeatedly warmed and refrozen over countless millennia and still survive their trips close to the Sun time and time again.

But Whipple overcame the uncertainties about a comet's physical structure by building on the much earlier work of Laplace and Bessel. Whipple's dirty snowball model proposed an icy nucleus that produced increasing quantities of gases by sublimation as the comet bathed in increasingly warm sunlight. The conglomerate part arose from the fact that the sublimation also released meteoric dust.

Whipple's model struck a successful chord because it suddenly explained a whole spectrum of what astronomers had observed in comets for decades. It seemingly explained how comets could produce large amounts of gas. It explained how jets and other structures could be observed near the nuclei of comets. It explained non-gravitational effects in comets that resulted from outflowing gas from the coma. It explained how sungrazing comets in the Kreutz group could survive close passages to the Sun. And it explained how comets produce meteor streams that in turn cause meteor showers in Earth's sky. Small details still puzzled Whipple and others, but the model he proposed caught on and nearly all astronomers believed in it.

It seemed that by mid-20th century the understanding of comets was coming together nicely, because it was also at about the same time that Jan Oort proposed his model for the huge shell of comets that surrounded the solar system, far from the Sun. The Oort Cloud hypothesis was also a long time in coming. In the early 20th century Swedish-Danish astronomer Svante Elis Strömgren (1870–1947) showed that hyperbolic orbits observed in comets must be due to gravitational perturbations by Jupiter. This meant that although comets originated from distant locales, they were not coming from interstellar space.

In 1932 Estonian astronomer Ernst Öpik (1893–1985) suggested that comets might be harbored in a distant cloud, which was stable somehow against the gravitational effects of passing stars. But it was Oort, in 1950, who actually studied the orbits of 19 comets and mathematically deduced the existence of the cloud on the basis of the semimajor axes – the half-lengths of the longest portions – of the comets' orbits.

Among the amazing inferences of Oort's work was that comets could remain in stable orbits to distances of about 200,000 astronomical units. (An astronomical unit is the distance between Earth and Sun, about 149.6 million km.) That incredible distance is some three-quarters of the way to the nearest star.

Some of these distant comets would, though, be gravitationally "kicked" inward by the influence of nearby stars. Over the 4.6-billion-year history of the solar system, Oort proposed, the orbital inclinations of these comets would have been totally randomized. And he further suggested that, in order to explain the number of new comets astronomers were discovering, the cloud of comets that would bear his name probably contains 200 billion comets. He also postulated that the total mass of the Oort Cloud would be about one-third the mass of Earth.

Along with his young student Maarten Schmidt (1929–) – who 13 years later would discover the first quasar – Oort studied the differences between "old" and "new" comets. The pair defined so-called new comets as fresh comets that were making their first appearance in the inner solar system, whereas old comets were returning for another trip around the Sun. New comets appeared to be richer in dust and brightened more slowly than older comets.

As research continued, support for the Oort Cloud model of comets only increased. The ability to produce refined, far more accurate orbits for comets increased throughout the 1970s, chiefly through the work of the English astronomer Brian Marsden (1937–2010) and Czech astronomer Zdeněk Sekanina. Not only has evidence for the cloud strengthened, but astronomers now believe that comets are gravitationally knocked into the inner solar system from the Oort Cloud by the overall gravity of the galaxy as a whole more so than by passing stars.

Oort believed in a vastly distant cloud of comets. But he also toyed with the idea of a second, closer disk of comets that could explain the replenishment of the Oort Cloud and potentially be another source of distant comets. In 1981 American astronomer J. G. Hills proposed in detail the existence of this inner cloud, which could extend inside 20,000 astronomical units – one-tenth the way to the limit of the Oort Cloud.

Most planetary scientists came to believe that Oort Cloud comets formed in the region between Jupiter and Neptune and then migrated outward into the cloud.

Astronomers now believe that as many as 2 trillion comets exist in the Oort Cloud. They also think that most surviving comets in the cloud formed in the region between Saturn and Uranus.

The science of comets seemed to be crystallizing quite nicely through the 1970s and early 1980s. And then the biggest buildup of hype, insanity, real science, and explosion of the amateur astronomy hobby our time has seen occurred. It all came riding along with the most recent appearance of Halley's Comet, which would take place in 1985 and 1986.

In their quest to study the most famous comet in history, planetary scientists planned multiple spacecraft missions that would encounter the celestial visitor, as well as readying their battery of ground-based telescopes. The intense study of Comet Halley (formal cataloged name: 1/P Halley) was so voluminous and the scientific results so important, that the moment became a dividing line in cometary science. The two resulting eras were simply "before Halley" and "after Halley," as if they were referring to the life of Jesus of Nazareth.

The appetizer in this great assault on a tiny, frozen block of ice occurred on September 11, 1985, when the *International Cometary Explorer* (*ICE*) spacecraft shot through the plasma tail of Comet 21/P Giacobini-Zinner. This comet has a nucleus some 2 km across and was discovered by French astronomer Michel Giacobini in 1900 and German astronomer Ernst Zinner in 1913. Although it wasn't armed with a camera, *ICE* measured particles, waves, fields, and plasma in the comet's tail en route to Halley. The probe passed within 8,000 km of Giacobini-Zinner's nucleus and confirmed astronomers' notions about the plasma tails of comets, measured ions in the tail, and measured a neutral electric current inside the tail.

The spacecraft missions to Halley heated up significantly in the spring of 1986, when five separate probes encountered the famous visitor. The first was the Soviet *Vega 1* probe, which flew past Halley on March 6, 1986. As it approached, *Vega 1* revealed two bright areas on the comet's nucleus. These turned out to be jets of material emanating from the comet. The craft showed Halley's nucleus was exceptionally dark and had a temperature of 300 to 400 K, warmer than expected. At closest approach, *Vega 1* whizzed just 8,889 km from Halley's nucleus as it snapped 500 images of the coma.

Second in line was the Japanese probe *Suisei* ("Comet") which encountered Halley on March 8 at a distance of 150,000 km. *Suisei* imaged Halley from such a large distance because its mission was to capture ultraviolet images of the comet's huge surrounding shell of hydrogen gas. The probe took up to 6 images per day of the comet and succeeded in its mission.

Third in the line of Halley probes was the Soviet *Vega 2*, which reached its closest approach on March 9. Sister craft of *Vega 1*, this probe (like *Vega 1*) encountered Venus first and then proceeded to the comet. As it approached Halley, *Vega 2* commenced by snapping 100 images and then unleashed a science suite of studying the physical parameters of the nucleus, shape, temperatures, and surface properties. Altogether, the probe captured 700 images with better resolution than its sister craft.

The next Halley probe was the Japanese *Sakigake* ("Pathfinder"), Japan's first interplanetary spacecraft. Encountering the comet on March 11, at the very great distance of 7 million km, the craft measured plasma wave spectra, solar wind ions, and interplanetary magnetic fields.

Finally, there was *Giotto*. The craft was named for Italian Renaissance painter Giotto di Bondone (1266/7–1337), who included Halley's Comet as the "Star of Bethlehem" in his *Adoration of the Magi*, which he painted in 1304–1306. Engineered by the European Space Agency, *Giotto* flew past Halley on March 14 at a distance of 596 km and was struck by small cometary particles in the process.

Giotto provided the best images of Halley, revealing its nucleus to be a coal-black, peanut-shaped chunk of ice measuring 15 by 10 km. About 10 percent of the comet's surface outgassed, producing the coma. The spacecraft's analysis of Halley's composition indicated the comet consists of 80 percent water ice, 10 percent carbon monoxide, 2.5 percent methane and ammonia, and a blend of hydrocarbons, iron, and sodium. Images revealed the comet was a miniature version of larger solar system bodies, showing small-scale features such as craters, ridges, and mountains. No outgassing was visible on the side of the comet aimed away from the Sun, only from the "daytime" side.

The first great round of spacecraft missions launched to a comet had revealed an enticing story of the basics of what makes up one of these icy bodies. Later flybys of other comets added a great deal to the story. On September 21, 2001, NASA's *Deep Space 1* probe whizzed past Comet 19/P Borrelly (Figure 1.2), an opportunity that produced much higher resolution images of a comet's nucleus than *Giotto*'s images of Halley. The spacecraft was something of a systems test mission, and, ironically enough, several of its functions failed. Despite this, the images of Borrelly depicted an incredibly dark cometary nucleus with several areas of outgassing.

Three years later, NASA's *Stardust* probe encountered Comet 81P/Wild 2 (Figure 1.3), a long-period comet that had been knocked into a short-period orbit by Jupiter. Here the mission was expanded into much more ambitious territory. Not only would *Stardust* study and image the comet from close range, but it also carried an aerogel collector and a return probe that would collect samples of the comet's dust and return them to Earth. (The collector also scooped up interstellar dust particles.) High-resolution images of the comet were also a primary objective.

On January 2, 2004, *Stardust* sped past Wild 2 at the relatively low velocity of 6.1 kilometers per second. The spacecraft was moving so slowly that the comet overtook it as the two orbited the Sun. The closest approach was 237 km, a little more distant than had been planned, as mission controllers became concerned over dust particle

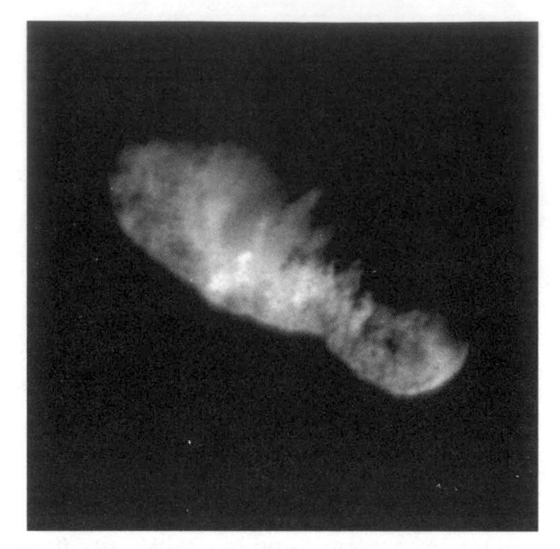

Figure 1.2. The nucleus of Comet 19P/Borrelly, appearing like a potato stretching 8 km across, appears before the camera on the *Deep Space 1* probe on September 21, 2001. The spacecraft encountered Borrelly at a distance of about 3,400 km, showing features as small as 45 meters per pixel. A variety of terrains and surface textures are visible, along with mountains and fault structures. Credit: NASA.

Figure 1.3. A wide-field shot of Comet 81P/Wild 2 taken March 21, 2010, shows a beautiful tail arcing sharply away from the coma. The imager used a 12-inch f/3.6 astrograph, a CCD camera, and stacked exposures. Credit: Gerald Rhemann.

collisions with the craft. Some two years later the sample return capsule separated from *Stardust* and reentered Earth's atmosphere, plummeting downward at 12.9 kilometers per second before deploying a parachute and slamming into the Utah desert.

Scientists published their initial findings on the returned samples by the end of 2006. They had a great deal to choose from; a million specks of dust had been

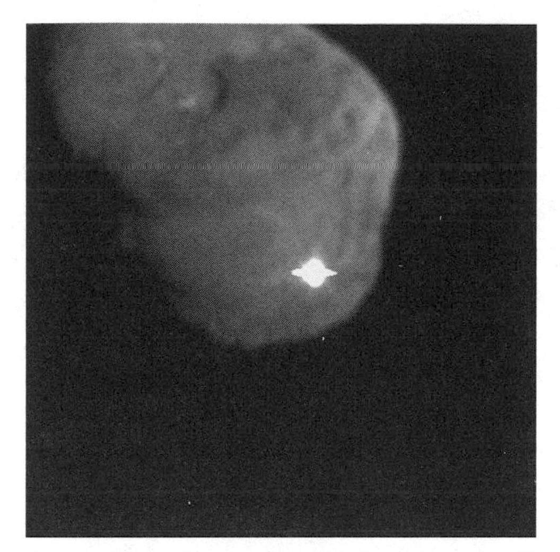

Figure 1.4. The moment of impact: a flash of energy and uplifted debris signals the collision of the *Deep Space 1* impactor into Comet 9P/Tempel 1 on July 4, 2005. Credit: NASA/JPL-Caltech/UMD.

deposited onto the collector's surface, and some 45 impacts from interstellar dust were found. Scientists announced they had identified a wide range of organic compounds in the sample, complex hydrocarbons, abundant silicates including pyroxene and olivine, some pure carbon, and methylamine and ethylamine.

Significantly, in 2011, researchers at the University of Arizona announced the discovery of iron and copper sulfide minerals in samples from Comet Wild 2, suggesting the formation of these minerals in liquid water. This was staggering, as previously no one had imagined that cometary nuclei could warm enough to melt a portion of their water ice. They also discovered some other secrets locked in the cometary grains.

Flybys have not been the only source of cometary drama in the solar system. In 2005 NASA launched the *Deep Impact* craft, a dual-purpose probe. Not only would *Deep Impact* approach Comet 9P/Tempel, a periodic Jupiter-family comet, but it would also strike right into the comet's nucleus, blasting material upward and studying the plume of debris. The aim was to answer fundamental questions about the nature of the nucleus, its composition, and perhaps even its origin.

The dramatic impact of *Deep Impact* was slated for July 4, 2005, and an enormous media buildup accompanied the mission. On April 25 the spacecraft opened the action by taking its first image of Tempel (Figure 1.4) at a distance of 64 million km. Sixty-nine days before impact, the probe spotted the comet with its medium resolution imaging camera and began an aggressive program of photographing the comet. Cameras recorded two periods of outburst from Tempel, on June 14 and June 22, and a week after the last outburst, controllers began to orient the craft for its strike.

Figure 1.5. On November 4, 2008, NASA's *EPOXI* spacecraft mission imaged the nucleus of Comet 103P/Hartley 2 at the time of its closest approach of some 694 km, revealing its pockmarked surface, lit by sunlight from the right side, and a cloud of tiny particles lit from the Sun. The comet's nucleus stretches just some 1.6 km across. Credit: NASA/JPL-Caltech/UMD.

Deep Impact released the impactor portion of the craft from the main probe and positioned to take a front seat in the comet's path - so the comet would in fact strike it. On the morning of July 4 - WHAM! The impact happened just as mission controllers expected it to, and images from the main probe showed a flash of light on impact. The impactor sent back pictures up until about 3 seconds before it struck. The energy released from the collision was equivalent to 5 tons of TNT, and the comet briefly lit up with a sixfold increase in brightness.

The science from *Deep Impact* surprised everyone involved. Scientists found the crater formed by the impact measured 100 meters across and 30 meters deep. The blast had liberated 5 million kg of water and 10 to 25 million kg of dust. The material consisted of much more dust and less water ice than scientists had thought they would find. The material was much finer grained than astronomers had guessed, consisting of particles akin to talcum powder rather than sand. Silicates and sodium were found in abundance but also clays and carbonates, suggesting the presence of liquid water for formation.

Astronomers likened the composition of the comet to a snow bank and suggested that as much as 75 percent of the space inside the comet was simply vacant. They concluded that Tempel 1 had formed in the icy outer solar system in the region of Saturn and Neptune.

Deep Impact was hardly finished, however. The craft received a second life in a dual-purpose mission that was dubbed *EPOXI*, short for *Extrasolar Planet Observation and Deep Impact Extended Investigation* - quite a mouthful. After several jugglings of a potential mission, the newly dubbed *EPOXI* swung past Comet 103P/Hartley (Figure 1.5), a small periodic comet of the Jupiter family, on November 4, 2010. Hartley thus became the fifth comet visited by a spacecraft and the smallest comet yet seen up close. The diameter of this potato-shaped block of ice is 1.2 to 1.6 km.

EPOXI's flyby showed that Comet Hartley was outgassing primarily carbon dioxide. The encounter distance of 700 km allowed the spacecraft's cameras to capture impressive images. Scientists found the "waist" of the peanut-shaped nucleus had been redeposited onto the comet. It also revealed the comet orbits along one axis but spins across another. And it revealed the comet's large ends contain relatively bright, blocklike objects as large as 16-story buildings.

Nor was *Stardust* through. Set for an extended mission and renamed *Stardust-NExT* (for New Exploration of Tempel 1), the orbiter that had previously collected particles would in 2007 redirect to Tempel 1. *Stardust* would now look at Tempel 1 to see what changes may have taken place to the comet since it was visited by *Deep Impact*. It would extend the mapping of Tempel 1, making it the most studied nucleus of a comet to date. And it would measure the mass and density of particles in the comet's coma.

The encounter date for *Stardust-NExT* and Tempel 1 was set for February 15, 2011. As it flew past the comet at a distance of 181 km, the spacecraft recorded 72 images. They revealed far more than *Deep Impact* showed, and the analysis of these images continues. As of this writing, one more comet flyby mission is in the works – the European spacecraft *Rosetta* mission, launched in 2004, will encounter and launch a landing probe onto Comet 67P/Churyumov-Gerasimenko on November 10, 2014. This will introduce yet another new era in the history of exploring comets with spacecraft.

Along with ground-based observations, space missions to comets have given astronomers a window into the early history of the solar system. Comets represent relatively pristine material from the formation of our Sun's family, and so studying them in detail sheds light on conditions dating back to the period when the Sun and its planets were forming. In subsequent chapters we'll explore what astronomers know about comets from their various means of study in much greater detail.

Before moving on, however, you ought to know about a somewhat sticky subject – the nomenclature of comets. How comets are designated and named comes about through a straightforward process, but one that has changed conventions over time (Figure 1.6). The body charged with the authority to name celestial objects is the International Astronomical Union (IAU), a worldwide group of professional astronomers created in 1919 and consisting of about 9,900 members.

The IAU's current comet naming guidelines were adopted in 2003. Many years ago, informal names were bandied about in all manner of ways. Not until the 20th century were comets systematically named for their discoverers. For decades, this system worked smoothly as typically one or two surnames neatly fit a given comet, as with Levy, Wilson-Harrington, or McNaught-Hartley.

Figure 1.6. Fragment C of Comet 73P/Schwassmann-Wachmann (Schwassmann-Wachmann 3) glowed brightly on June 1, 2006, when it was captured with an 8-inch f/3.8 astrograph, a CCD camera, and stacked exposures. Credit: Gerald Rhemann and Michael Jäger.

Recently, however, the world has sprouted networks of search telescopes designed specifically for finding comets and asteroids against the stellar background. These wide-field charge-coupled device (CCD) surveys have turned up numerous comets and – aside from making the probability of a human discovering a comet even more challenging – have changed the way comets are named. Now there are, or are likely to be, numerous examples of LINEARs, ISONs, NEATs, PANSTARRS, and others, named for the search networks.

Following the suspected discovery of a comet, astronomers around the world know to alert the IAU's Central Bureau for Astronomical Telegrams, at Harvard University in Cambridge, Massachusetts. The Central Bureau, or CBAT, was founded in Kiel, Germany, in the 1880s as the world's first clearinghouse for astronomical observations and migrated to Copenhagen, Denmark during World War I and finally Cambridge in 1965. Three important astronomers have directed the CBAT since its move to the United States, and they are familiar to all astronomers – Owen Gingerich

(1930–), director from 1965 to 1968; Brian Marsden, director from 1968 to 2000; and Daniel W. E. Green, director from 2000 to the present.

Not only does the CBAT alert other observatories so they can start observing a suspected comet right away (and look for observations they might have already captured), but it rides herd over the calculation of an orbit and other parameters such as positions, magnitudes, sizes, and other observational data. The CBAT issues these late-breaking observations on the famous *IAU Circulars*, both on printed card-sized alerts and in electronic form.

Despite the many changes over recent years, comets are still named, if possible, with the surnames of their discoverers. The CBAT prefers to limit the name to two independent discoverers, if appropriate, although in the past three or even more names have sometimes been used. Chronology rules – the person who found the comet first gets his or her name in the first spot. And of course search network names work the same way. In rare cases extremely bright sungrazing comets have become visible suddenly to a large number of people all at once, and in these cases the CBAT has adopted the phrase "Great Comet" or a similar designation rather than using discoverers' names.

All these rules apply to comets in modern times. In the old days, comets were usually named after the year in which they were found. So we had the "Great Comet of 1680," the "Great September Comet of 1882," and so on. For a time, after the discovery that some comets were periodic, only the periodic comets were named for their discoverers. So we had Halley's Comet, Encke's Comet, and Biela's Comet, while comets that appeared just once were still identified by the year of their appearance.

The naming of comets is one thing, the designations another. Before 1994, comets were given a provisional designation based on the year of their discovery and followed by a lowercase letter indicating the order of discovery within a year (a, b, c, etc.). So Comet West (originally 1975n) was the 14th comet discovered in 1975. Additionally, once a comet was observed thoroughly, through its perihelion passage, it was given a permanent designation consisting of its perihelion year followed by a Roman numeral indicating the order of its perihelion passage for the year. So Comet West became 1976 VI, the sixth comet to reach perihelion in 1976.

Because of the cumbersome nature of this system, in 1994 the IAU changed the naming conventions. The system is now more systematic and also creates several categories of comets that help classify them, as the number of comets discovered steadily increases. Comets are designated with the discovery year followed by a letter indicating the half-month of discovery and a number indicating the order of discovery. So the first comet discovered in the first half of January 2014 would be designated 2014 A1.

Additionally, classifications of comet types have a set of prefixes attached. P/ indicates a periodic comet, a comet that has been observed at more than one perihelion and with a period of less than 30 years. C/ indicates a nonperiodic comet. X/ indicates a historical comet for which no reliable orbit could be calculated. D/ indicates a periodic comet that broke apart or has been lost. A/ indicates an object that was first identified as a comet but is actually an asteroid.

It takes a bit of getting used to, but the system of nomenclature in place over the past 20 years works well. Only a few mysterious flies are in the ointment; just a handful of bodies in the solar system are classified as both comets and asteroids and stand forth as examples of the increasing ambiguity between the small body types. These include 2060 Chiron (95P/Chiron), 4015 Wilson-Harrington (107P/Wilson-Harrington), 7968 Elst-Pizarro (133P/Elst-Pizarro), 60558 Echeclus (174P/Echeclus), and 118401 LINEAR (176P/LINEAR). More to come on these later.

So I had come a long way from my first look at Comet West in the Ohio backyard, opening a doorway to the icy, distant past of the solar system. What I didn't appreciate at first, however, was that I had opened a special door – one that led to a world of what astronomers like to call Great Comets.

2

Great Comets

There are comets and then there are comets. I didn't know it when I caught my first glimpse of Comet West back in the spring of 1976. As I gazed at the brilliant nucleus and twin tails of that magnificent visitor, I was looking at a Great Comet. Astronomers like to use the term "Great Comet" for exceptionally bright comets, but there is no official classification as such.

Nonetheless, both in the minds of amateur astronomers and in the historical record, it's pretty clear when a comet is great. You know a Great Comet when you see one. It's a brilliant, naked-eye spectacle that causes the most casual viewer, your neighbor Fred or your Aunt Martha, who has never really observed the sky, to look up at the glowing specter and say, "WOW!" – or maybe even something a little stronger.

A number of factors influence how bright and well developed a comet will become. First and foremost is how close the comet will be to the Sun at perihelion. The closer the comet is to the Sun, the brighter it will be, as it will warm and outgas significantly more as it is closer to the Sun. Because of the inverse square law, normal objects are only one-quarter as bright when they are twice as far away from the Sun.

But comets are somewhat different. They not only reflect sunlight as other bodies do, but they stream gas and dust that can reflect sunlight over a larger area. And they also can fluoresce – that is, glow of their own accord as photons are produced during the process of heating and sublimation. The brightness of a comet is close to the inverse cube of its distance from the Sun; that is, if a comet is twice as close to the Sun as its twin, it will be about *eight* times brighter. Predictably, then, many Kreutz Sungrazers tend to be bright comets.

Another major factor in the making of a Great Comet is how close it passes to Earth. In 1983 the diminutive Comet IRAS-Araki-Alcock (C/1983 H1) passed closer to Earth than any other comet in the last 200 years. Had it not whizzed by us at a distance of just 4.6 million km (just 12 times farther away than the Moon), the fuzz-ball would have been an ordinary telescopic comet. The circular cloud was an easy naked-eye object and appeared as large as the Full Moon and moved over 30° of sky per day, so that you could watch it shift position against the starry background dramatically during a single night.

Although IRAS-Araki-Alcock was a naked-eye object, it wasn't bright enough to be considered a Great Comet. Thirteen years later, however, the intrinsically faint Comet Hyakutake (C/1996 B2) flew past Earth at the slim distance of 15 million km. What otherwise would have been a modest naked-eye comet wowed observers all over the globe, shining brightly at magnitude 0 and with its tail stretching an incredible 80° across the sky. It made for a stunning sight, even for beginning sky watchers.

The third factor dictating how well a comet will brighten is the physical nature of the comet itself. The size of the nucleus is one consideration: Cometary nuclei vary from a few meters to several dozen kilometers across. The wild card that prevents astronomers from accurately predicting their brightness before they move in close is the physical nature of the nuclei. All comets are different – densities, composition, amount of dust, "freshness" of the ices, distribution of material within the nucleus, and how and when the nucleus becomes active with outgassing and releasing particles of dust.

The experience of a comet in terms of inner solar system visits is also important. A fresh comet that has never seen the inner solar system will brighten far more dramatically than an older comet that has made many trips inward, warming and refreezing repeatedly, and depleting its stockpile of volatiles.

And unpredictable outbursts do occur. On October 23/24, 2007, Comet 17P/Holmes (Figure 2.1) exploded in a historic outburst, increasing in brightness by a factor of 0.5 million. In a mere 42 hours it brightened from magnitude 17 to 2.8, becoming a naked-eye object literally overnight. With a nucleus just 3.4 km across, Holmes temporarily became the largest object in the solar system, its immense outgassing creating a vast halo of particles.

The truly Great Comets belong to an exclusive club. Over the past 2,000 years of recorded observations, humans have seen around 70 Great Comets. The following chapter samples two dozen of the most intriguing. And the record begins with a story that centers on one of history's most colorful characters, the Roman dictator Julius Caesar (100–44 B.C.).

In antiquity, comets were often associated with great events – good or bad. One of the most famous examples followed the death of Caesar on March 15, 44 B.C. After Caesar's assassination on the Ides of March, lower- and middle-class Romans became enraged that a small group of the elite had killed their leader – dictator though he

Figure 2.1. During its return to the inner solar system in October 2007, Comet 17P/ Holmes explosively brightened by a factor of a half million. The huge, ethereal coma, bright nucleus, and fan-shaped tail produced a highly memorable appearance. This image was made December 2, 2007, using an 80 mm refractor and 60 minutes of exposure time. Credit: Chris Schur.

was. The appearance of a brilliant comet over the skies of Rome soon after Caesar's death was taken by many Romans as the deification of the fallen leader.

The Great Comet of 44 B.C. (C/–43 K1) became known as Caesar's Comet and was one of the most brilliant and celebrated comets of the ancient world. The comet may, in fact, have been the brightest daytime comet recorded in history, and it is one of only five comets to have a negative absolute magnitude, calculated at –4.0. (Absolute magnitude is a measure of an object's intrinsic brightness, independent of its distance.) Greek sources also reported the comet's appearance, and records from Chinese and Korean observers of a bright comet may also coincide with Caesar's Comet.

The earliest surviving reference to the comet is from Augustus (63 B.C.– A.D.14), Caesar's great-nephew, adoptive son, and successor – and founder of the Roman Empire – who used the comet's appearance as propaganda for his reign. In ancient records the comet was known as Sidus Iulium ("Julian Star") or Caesaris Astrum ("Caesar's Star"). Visible in broad daylight, the comet allegedly appeared suddenly during a festival of sporting games held in Caesar's honor. The event probably took place in July of that year, some 4 months after Caesar's death.

In his celebrated *Natural History*, the Roman author and philosopher Pliny the Elder (23–79) quoted Augustus himself. "On the very days of my games," wrote Augustus, "a comet was visible over the course of seven days, in the northern region of the heavens. It rose at about the eleventh hour of the day and was bright and plainly seen from all lands. The common people believed that this star signified the soul of Caesar had been received among the spirits of the immortal gods."

The date of the comet's appearance has long been the subject of uncertainty. Some Roman writers mentioned the comet's appearance "during the games of the Venus Genetrix," and these were normally held in September. However, at this time special games were created, the Iudi Victoriae Caesaris, and these events were held in late July. So recent astronomical historians have concluded the comet was probably seen during late July 44 B.C.

The alleged Chinese observations of this comet are harder to pin down. In the *Book of Han*, published in 100, the Chinese historian Ban Gu (32–92) reported that a "broom star" appeared during the summer of 44 B.C., about May 18 through June 16. The comet appeared in the group of stars known as Shen, which included many of the brightest stars in Orion. The account reported the comet was visible in the northwest and was "reddish-yellow and measured about 8° long. After several days passed it measured over 10° and pointed toward the northeast."

Korean accounts also mentioned a comet amid the bright stars of Orion, a "sparkling star" seen at the same time as the Chinese observations. But in his landmark work *Cometography*, the comet expert Gary Kronk suggests the Korean accounts were probably copied from the Chinese writings.

The identities of the Chinese and Korean accounts are not exactly clear. In 1783 the French astronomer Alexandre Pingré (1711–1796) suggested the two accounts referred to different comets, and a debate has raged, on again, off again, ever since. In their voluminous work on Caesar's Comet published in 1997, the historians John T. Ramsey and A. Lewis Licht argued that the Roman comet was visible during July, and that the Roman and Chinese comets were indeed one and the same. They proposed the comet was bright when viewed in China, faded, and then underwent a dramatic outburst that enabled it to be seen as a brilliant object in Rome.

Fast-forwarding more than a millennium takes us to another of history's Great Comets. When Chinese observers spotted a "broom star" on April 2, 1066, they could not have known that it would become the most celebrated comet in history. The first observations of what would come to be known as Halley's Comet (Figure 2.2) occurred with the fuzzy star in the constellation Pegasus, in the eastern morning sky, and with a tail about 7° long. The comet sank progressively in the east, closer to the Sun.

The comet's name association with Edmond Halley would of course begin much later; Halley himself wouldn't be born for another 590 years; he wouldn't recognize the comet's periodicity for yet another 49 years; and the comet wouldn't return again for 53 years beyond that, thereafter becoming known as Halley's Comet. Still, the apparition of 1066 – the first round of observations of what became Comet Halley – is one of the grandest early events in the history of cometary astronomy.

Chinese observations of the first recorded appearance of Halley's Comet were voluminous. Four lengthy accounts provide details and reveal that the comet

Figure 2.2. The world's most famous comet, 1P/Halley, puts on a show for observers as it floats among the stars of the summer Milky Way on March 21,1986. Credit: ESO.

reappeared in the evening sky on April 24. Depending on which account is to be believed, some suggest the comet stretched 15° long and others claim it had virtually no tail. The *Sung shih*, written in 1345, claims the comet moved through Draco, Ursa Minor, Cepheus, and Camelopardalis and eventually joined the stars of the Big Dipper. On April 25, the text says, the comet "retained its rays and measured over 10° in length and 3° in breadth."

Chinese accounts suggest the comet continued moving through the stars of Auriga, Taurus, and Gemini, and by the time it scurried into Virgo, it spanned 15° and showed a "broomlike vapor." The *Sung shih* describes the last date of observation as June 7, when the would-be Halley's Comet was in the southern constellation Hydra, and that it had remained visible for 67 days.

Japanese and Korean observers also recorded the comet's apparition. Published in 1715, the Japanese *Dainihonshi* reported the "broom star" appeared in the eastern morning sky on April 3, 1066. Japanese observers recorded the comet's tail as 7° long and stated the comet remained visible for 20 days in the morning sky before sliding over to the evening sky. Published in 1451, the Korean *Koryo-sa* recorded a "star" visible on April 19, 1066. The object was "like a moon [that] rose from the northwest. Presently it transformed into a comet."

The 1066 apparition of Halley's Comet made a big splash among European observers. Italian observers spotted the object in both the morning and evening skies. English viewers had a field day. The historian Henry of Huntington (*ca.* 1088–*ca.* 1154) penned rather grandly, "Thus the hand of the Lord brought to pass the change which a remarkable comet had foreshadowed in the beginning of the same year; as it was said, 'In the year 1066, all England was alarmed by a flaming comet.'"

Henry was referring, after the fact, to the many interpretations of the comet as some sort of strange omen – an idea that received apparent confirmation later in the year when King Harold II of England (*ca.* 1022–1066) died in the Battle of Hastings. It could have been taken as a good omen, conversely, by William the Conqueror (*ca.* 1028–1087), who defeated Harold and rose to become the first Norman king of England.

The presumed connection of the comet to the battle and Harold's downfall was captured magnificently in the Bayeux Tapestry, an embroidered cloth 230 feet long that depicts events leading up to the Norman conquest of England. Its origins are not clear; the earliest known reference to this famous work of art was in a 1476 inventory of goods at the Bayeux Cathedral. The work can now be seen, depicting Halley's Comet as a flaming star with a tail, in the Musée de la Tapisserie de Bayeux in Bayeux, Normandy, France.

Observers in France, Germany, Belgium, and Ireland wrote glowingly of the comet. Published in 1220, the French text *Annales Sanctae Columbae Senonensis* proclaimed, "In this year with the shining sun in the first part of Taurus, on April 24 there appeared a comet in the last part of the same constellation, which was scattering sulfurous fires toward the south. When it began it extended as far as Saturn, also situated in the last part of Gemini, and from its very rapid motion it was understood to have been Mercury. However, after 15 days when the Moon was gleaming and coming close to it, it was soon extinguished."

Russian accounts mention this comet in a variety of places. The *Nikonian Chronicle* of 1520 mentions a "very large star with blood-colored rays" that "rose in the evening after sunset and remained in the sky for seven days." Published in 1116, the *Primary Chronicle* records "a portent [appeared] in the west in the form of an exceedingly large star with bloody rays, which rose out of the west after sunset. It was visible for a week and appeared with no good presage."

Published in Egypt in the 15th century, *History of the Patriarchs of the Egyptian Church* is a work that was compiled for more than 400 years. This work refers to the 1066 showing of Comet Halley: "It was a great star like the moon on the night of its fullness, and it appeared also the second day where the sun would be during the 8th hour of the day, and its locks blazed such that it became like a lantern when the light is lit in it." Importantly, this account suggests daytime visibility.

In Armenia, the historian Aristakes Lastivertsi (1002–1080) wrote about the comet in his famous *History: About the Sufferings Visited Upon by Foreign Peoples Living around Us*. His superbly written account follows:

> For while it was fully lit, it was in its mid-course, speedily headed toward earth a shadowless hemisphere, wearing an expansive robe woven of sins committed over a long period. But that robe which it had donned, so thickly enveloped it that it blocked those unbelievably brilliant rays. And [the brightness] which had been so strong that the eye

> could not gaze at it, then became weaker than the [distant] stars and merely its outline was visible.

It was the English astronomer John Russell Hind (1823–1895) who in 1850 calculated a relatively precise orbit for this comet and suggested it was a previous apparition of Comet Halley. Noting some differences in the orbital elements for the Great Comet of 1066 and Halley, Hind suggested, "In the lapse of so many centuries, may not the planetary perturbations have produced alterations in the elements at least equivalent to those here exhibited…?"

But Hind also correctly pointed out that early observations were so confused that any differences between the orbit of the 1066 comet and that of Halley could be meaningless, and that the two comets were indeed one and the same. It turned out he was right.

Fifty years after the 1066 appearance of Halley's Comet, another Great Comet flashed across Earth's skies. Relatively little of definitive value is known about the Great Comet of 1106 (X/1106 C1) because not enough precise observations exist to calculate its orbit. But clearly a spectacularly bright comet was visible in this year and observers noted it in Europe and Asia. On the basis of the data that do exist, Brian Marsden proposed this comet must have been a member of the Kreutz Sungrazer family.

The original account of this comet that survives appears in a work by the Belgian historian Sigebertus Gemblacensis (*ca.* 1030–1112). His *Chronica* reports that a daytime "star" appeared in February 1106 that was placed approximately in the constellation Aquarius and remained visible throughout the whole month. The anonymous work *De Significantione Cometarum* reports on a comet viewed from Palestine in February 1106, which was found in Pisces and had a tail extending toward Gemini that extended 100° across the sky.

Japanese, Korean, and Chinese accounts all mention the comet. Various Chinese texts mention that it was first seen on February 10, as a "broom star" that appeared in the west and showed rays that scattered in all directions across the sky. Chinese observers reported the tail as 60° long and 3° wide, spanning from Andromeda through Aries and into Taurus.

Six days later the European map of observers lights up with myriad accounts. English, Scottish, French, Italian, Dutch, Belgian, and German observers all left reports. In England, authors of the *Chronicon ex Chronica* described the comet thus:

> On Friday, in the first week of Lent, the fourteenth of the Kalendas of March [February 16], in the evening, a strange star was visible between the south and west, and shone for twenty-five days in the same form and at the same hour. It appeared small and dim, but the light which issued from it was exceedingly clear; and flashes of light, like bright

> beams, darted into the star itself from the east and the north. Many affirmed that they saw several strange stars at this time.

Most accounts reported seeing the Great Comet of 1106 for a period of 15 to 70 days. The discovery date has typically been given as February 16. In more recent times, several astronomers have suggested this comet may have represented a previous appearance of another wonderful object, the Great Comet of 1680.

The High Middle Ages of the 13th century brought another visitor, the Great Comet of 1264 (C/1264 N1). Discovered on July 17, the comet was visible in the evening sky and was moving toward a postperihelion close approach with Earth of only 27.4 million km. Japanese observers discovered the comet in the northwest, among the stars of Cancer. European observers also saw the comet in the evening sky. The best viewing would happen as the object swung over into the morning sky.

Chinese sky watchers reported that tail length was more than 100° for the Great Comet of 1264, and that it only became invisible when the Sun was high in the sky. The comet moved into Gemini and then on southward into a position between Canis Major and Orion. By late August Korean astronomers wrote that the comet's tail had increased in length and the separate portions of the tail had reunited. By October the comet was becoming more difficult to see; the last observation had been made by Japanese observers on October 9. The comet may have been visible for as long as 4 months. The Italian manuscript *Chronica*, published in 1288, reported of the comet: On the seventh of August, 1264 ... a strange comet appeared, which nobody living had ever seen before. Rising in the east with great brightness, it traveled to the west with an extremely bright tail. And although it signified many different things in various parts of the world, this one event for certain it signified: for it appeared at the time that Pope Urban began to grow ill, lasted for more than three months, and on the very night the Pope died, it disappeared.

Nearly 150 years passed before the appearance of our next object, the Great Comet of 1402 (C/1402 D1). Discovered February 8, 1402, the comet was visible for 3 months and would swing past Earth at a distance of 105.6 million km on February 20. During March 1402 this object became visible in broad daylight for 8 days, the longest period of daylight visibility of any comet.

The Renaissance humanist Giacomo di Scarperia, known as Jacobus Angelus, wrote an account of the comet in the 15th-century *Tractatus de Cometis*. Scarperia wrote that

> around the beginning of February a comet appeared here in Swabia for many days. We first saw it in Ulm the fifteenth of March. On the [22d] day it was toward the north of west, and it set crossing the horizon

Figure 2.3. Comet Machholz (C/2004 Q2) shows a puffy coma and thin, sharply defined tail in this image made December 12, 2004, with a 76 mm refractor at f/6, a CCD camera, and stacked exposures. Credit: Jose Suro.

> at the point where the sun sets when it occupies the summer solstice point. Its size was rather greater than that of Venus when it becomes visible in our hemisphere before sunrise, but not as bright. Its color was the color of Venus, which is rather metallic.

Another 150 years passed, moving into the age of the rise of the West, Copernicus's proposal of the heliocentric theory of the universe, and the exploration of the high seas. Now two exceptionally bright comets appeared in succession, separated by only 21 years.

The Great Comet of 1556 (C/1556 D1) was discovered on February 28 and made a close passage to Earth of only 12.5 million km on March 12. German astronomer Joachim Heller (*ca.* 1518–1590) was the first to see the comet, spotting it in the constellation Virgo. Asian astronomers independently found the comet, as Chinese and Korean observers noted its appearance on March 1. Chinese astronomers typically described it as a "broom star" and noted a southwestward-pointing tail about 1° long.

As this Great Comet moved rapidly northward, Heller kept observing it and by March 6 described its tail as "of mingled pale and dingy colour, such as the flame of burning sulfur usually presents." Back in China, while traveling, Portuguese Dominican friar Gaspar da Cruz (*ca.* 1520–1570), the first to write detailed European accounts of China, described the comet's appearance: "A comet appeared in a star in the northern hemisphere," he wrote, "which was visible in all regions in India and in Portugal. And it appeared almost for the space of fifteen days, likewise having been seen in China."

Despite the fact that Edmond Halley calculated a refined orbit for this comet in 1705, later astronomers thrust the Great Comet of 1556 into some controversy because they believed it may have been a reappearance of the Great Comet of 1264. In 1856 and for several years following, however, Dutch astronomer Martin Hoek (1834–1873) reanalyzed and computed a number of orbits and concluded the two comets were not associated with each other. After 5 years of argument, the debate was settled.

Merely 21 years after C/1556 D1 graced Earth's skies, an important object appeared – the Great Comet of 1577 (C/1577 V1). No one could have known it at the outset, but this object would become the first comet that was proven to exist outside Earth's atmosphere, putting comets in their proper perspective in the solar system. Many different astronomers and writers commented on the discovery of this comet, with conflicting details. French astronomer Alexandre Pingré (1711–1796) later described the comet as having been discovered in Peru on November 1. Mexican sources reported sightings of a brilliant comet on November 4 and 6. Correcting local times to Universal Time, astronomers settled on the Peruvian account and a discovery date of November 2.

The great Danish astronomer and nobleman Tycho Brahe (1546–1601) observed this comet at great length, providing numerous accounts of its appearance. He described sailors in the Baltic Sea observing the comet on November 9. Two days later he wrote an account based on the observations of others:

> This new birth in the heavens revealed itself, namely a comet with a very long tail, and the body of the star was whitish, though not with the bright gleam of the fixed stars but somewhat darkish, almost like the star Saturn in appearance, which indeed at that time stood not far away from it. Its tail was great and long, curved over itself somewhat in the middle, of a burning reddish dark color like a flame penetrating through smoke.

On November 13 Tycho first saw the comet for himself. His account reads, "I perceived in [the west] a certain bright star, which appeared as distinct as Venus, when near to the Earth and when seen before sunset or after sunrise. For the rays or chevelure of the star could not yet be perceived, the sun, still above the horizon, entirely obliterated the feeble brightness of its rays." In the dark sky that followed, Tycho noted a yellowish coma 8 arcminutes across and a tail stretching more than 21° long and spanning 2.5° wide.

The Great Comet of 1577 made its closest approach to Earth on November 10 at a distance of 93.8 million km and traversed a path that carried it from Lupus through Scorpius, Ophiuchus, Sagittarius, Delphinus, and Pegasus. By December the tail had shrunk to 7° and after the New Year few observations were made. On January 27

Tycho described the comet as having "grown so small that one could hardly see it, and for aught I know, it faded soon after, and was gone."

Tycho then wrote extensively about the Great Comet of 1577 and drew some scientific conclusions from his observations. He claimed the raylike structure of the comet's tail represented the Sun's rays shining through the comet. He tried to calculate the parallax and determined the comet was at least 230 times the radius of Earth away from our planet. He also realized that his size measurements of the comet – combined with the distance he had calculated – meant the comet was much larger than anything anyone had previously imagined. He believed that the coma spanned about a quarter of the diameter of Earth and that the tail was millions of kilometers long, which at the time was shocking.

Comet C/1577 V1 was a milestone for planetary astronomy. Another century would pass before our next Great Comet would become visible, however. This historic mark pushed past a technological one: Galileo invented his telescope in 1609 and first used it extensively as an astronomical instrument. Now in the telescopic era, German astronomer Gottfried Kirch (1639–1710) made the first telescopic discovery of a comet on the morning of November 14, 1680.

Kirch was observing a waning crescent Moon and Mars, which lay nearby, with his telescope. Positioned near the Moon, the field of view swept up a star that he stopped and looked at, and that he found was not listed in the famous star catalog of Tycho Brahe. He moved a short distance from the star and suddenly saw "a sort of nebulous spot, of an uncommon appearance." He thought the object was a comet or a "nebulous star" that resembled the one in the "girdle of Andromeda," that is, the Andromeda Galaxy (M31). He observed no tail.

The object Kirch found turned out to be the Great Comet of 1680 (C/1680 V1), which came to be known as Kirch's Comet. It is this Great Comet that may be associated with, or at least has a similar orbit to, the current Comet ISON (C/2012 S1). Kirch's Comet became one of the most brilliant comets of history, passing 73.2 million km from Earth on January 4, 1681. The comet traversed a path from Leo through Virgo, Libra, Scorpius, Sagittarius, Scutum, Pegasus, Andromeda, and Perseus before it faded in February 1681.

Because the comet was a sungrazer that was rapidly approaching both Earth and the Sun, it brightened to naked-eye visibility soon after Kirch's discovery. Numerous observers spread across the globe observed the comet in November, from the Philippines to China to the United States to England. America's first colonial astronomer, Arthur Storer (*ca.* 1648–1686), spotted the comet on November 29 and estimated the tail length as between 15° and 20°.

In December, as it moved through Scutum, the comet emerged into the morning sky, where observers in England and France tracked its movements. The comet's reemergence was dramatic; many observers described its long, narrow tail. In

England, John Flamsteed (1646–1719), Britain's first Astronomer Royal, described the comet as a beam of light about the width of the Moon, extending straight up from the horizon.

In France, the Jesuit Jean de Fontaney (1643–1710) of the College of Clermont observed the comet on December 28 and noted, through a 12-foot telescope, it as "differing both from stars and planets, being dusky light like a cloud, about the size of the moon, and brighter in the middle than the extremes."

After the New Year, the comet began to fade slightly. On January 5 Flamsteed compared the nucleus to a 3d-magnitude star. Reports of the tail length during this period had it ranging from 50° to 75° long. On January 15 Isaac Newton (1642–1726), perhaps the greatest scientist who ever lived, spied the comet and thought the tail stretched 40° and was "curved, and the convex side thereof lay to the south."

During February 1681 the comet faded from view for most observers. The comet shrank to an object with a tail 6° or 7° long with a nucleus merely as bright as a 7th-magnitude star. March was a period of total inactivity during which only the indefatigable Newton continued his observations. Finally, on March 20, Newton made his last observation and reported the comet as "barely discernable."

Another Great Comet appeared in 1744, as colonial America was awakening, Catherine the Great ruled over Russia, France declared war on Great Britain, and the First Saudi State was founded by Mohammed Ibn Saud. The Great Comet of 1744 (C/1743 X1) was one of the most beautiful and brightest comets in history and for a time was visible in daytime skies.

Two observers first spotted this comet - Dutch amateur astronomer Dirk Klinkenberg (1709–1799), whose daytime job was as secretary of the Dutch government, and Swiss-French astronomer Jean-Phillippe de Chéseaux (1718–1751). The respective discovery dates were December 9 and 13, 1743. This comet looped through Triangulum, Aries, Pisces, Andromeda, Pegasus, Aquarius, and Cetus over the 4 months that followed, making its closest approach to Earth of 123.7 million km on February 26, 1744.

The comet was widely observed by many who thought they each had discovered it, and by late January 1744 it became spectacular. German astronomer Gottfried Heinsius (1709–1769) observed the comet on January 25, displaying a triangular ray situated with the tip near the comet's nucleus, and remarked, "The lateral borders of this needle seemed curved as if they were pushed from the inside to the outside by the action of the sun."

In February the comet continued to grow in brightness. English observer George Smith (1700–1773) described it as equaling "any Star of the first Magnitude, except Sirius"; estimated its tail length as more than 25°; and called it "superior to Jupiter in lustre." On February 25, Smith wrote, "I compared it with Venus, and found it very little short."

During March the comet sank into the morning twilight, hiding the nucleus but allowing the prominent tail to remain sticking up above the horizon. And the tail wasn't just an average tail: Many observers marveled at the comet's multiple tail structure, and de Chéseaux famously drew six tails extending above the horizon, which he published in beautiful engravings the same year.

Planet Earth welcomed the 19th century before the arrival of our next Great Comet. Discovered by French astronomer Honoré Flaugergues (1755–1835) on March 26, 1811, the Great Comet of 1811 (C/1811 F1) turned out to be one of the most spectacular in history. The comet rose to naked-eye visibility by April and remained watchable by the eye alone for more than 8 months. Telescopic observers followed the comet for nearly a year and a half.

This comet had a peculiar historical ring, even in its day. French emperor Napoleon I believed the comet's apparition foretold momentous success for the invasion of Eastern Europe and Russia he planned. For astronomers, however, the comet was simply a terrifically bright object. Flaugergues found the comet in the now-antiquated constellation Argo Navis, and moonlight temporarily interrupted observations during early April. Soon thereafter the comet delighted sky watchers with its brilliance. English naturalist William Burchell (1781–1863) observed the comet from Cape Town, South Africa, where in June an earthquake struck. He recorded that locals associated the comet with the earthquake and "drew from this two-fold portentous sign, the certain prognostics of the annihilation of the Cape."

The comet slipped too close to the Sun in late June and reemerged in the mid-August evening sky. On September 1, Canadian fur trader Alexander Ross (1783–1856) saw the comet during the expedition down the Columbia River in Oregon that was financed by John Jacob Astor. He saw "about 20 degrees above the horizon, and almost due west, a very brilliant comet, with a tail about 10° long. The Indians at once said it was placed there by the Good Spirit – which they called Skom-malt-squisses – to announce to them the glad tidings of our arrival; and the omen impressed them with a reverential awe for us, implying that we had been sent to them by the Good Spirit, or Great Mother of Life."

Throughout the autumn, the great German-English observer William Herschel (1738–1822), discoverer of the planet Uranus, repeatedly viewed the comet and made lengthy notes about its appearance. On September 18 he remarked that "the appearance of the nebulosity … perfectly resembled the milky nebulosity of the nebula in the constellation of Orion, in places where the brightness of the one was equal to that of the other." Of the 17°-long tail on October 12, he penned, "the two streams remained sufficiently condensed in their diverging course to be distinguished for a length of about six degrees, after which their scattered light began to be pretty equally spread over the tail."

The Great Comet of 1811 dazzled observers for several more months. Another Great Comet would be widely visible in March, this one 32 years later. The first mention of a comet sighting in February 1843 reportedly was in a New York newspaper, and the account placed the comet in Cetus. Comet expert Gary Kronk, in *Cometography*, analyzed this and other early reports and fixed on February 6, 1843, as the discovery date for what became the Great Comet of 1843 (C/1843 D1), often called the Great March Comet.

This comet would rise to become a sensational naked-eye object, swinging past Earth at a distance of 126 million km on March 5. It moved from Cetus through Aquarius, Eridanus, and Orion and was visible for a time in broad daylight. The comet was visible in the evening in February before slipping into the Sun's glare, but observers caught it at month's end just a short distance away from the Sun. On February 28, in fact, the comet's nucleus transited the Sun's disk, passing very close to its center.

The Great Comet of 1843 was so brilliant that numerous observers saw it in the daytime. At the end of February Italian observers described a "very beautiful star" with a tail stretching 5° long. Although Europe was in a cloudy period, North American sky watchers gobbled up the comet. Reports suggested that on February 28, a "large part of the adult population" of Waterbury, Connecticut, observed the comet.

As March 1843 began, the Great Comet's tail moved into the evening sky. On March 3 a number of observers reported seeing a double tail. On this date Charles Piazzi Smyth (1819–1900), who would become Scotland's Astronomer Royal, viewed the comet from the Cape of Good Hope, South Africa. He wrote, "To the naked eye there appeared a double tail, about 25° in length, the two streamers making with each other an angle of about 15°, and proceeding from the head in perfectly straight lines."

On March 6, Smyth penned, "the nucleus is the broadest part of that end of the comet; all rays come from the posterior side, and are pretty equal in brightness, with the exception of a narrow bright streak in the middle, which runs for about 3° or 4° along the middle of the tail, and then verges to the north side."

By mid-March, the comet was still bright, but recorded observations are fewer. In Nice, France, Edward Cooper described it as "a long white light near the western horizon which had somewhat the appearance of that kind of cloud commonly called cirrostratus. This I conceived it to be, although there were very few clouds in the sky at the time."

The comet was widely visible throughout the rest of March and by mid-April had faded into oblivion. Smyth made the last observation of this Great Comet on April 20 from the Cape of Good Hope, calling it then "of the last degree of faintness." Years later, in 1904, Virginia abolitionist Moncure Conway (1832–1907) wrote of the year 1843 that the greatest sensation had been caused by the comet. "There was a

Figure 2.4. Comet McNaught (C/2006 P1) shines brightly over Argentina's Patagonia region in this image made January 18, 2007, with a 54 mm lens at f/5.6 and a 13-second exposure. Credit: Martin Moline.

widespread panic," he recalled, "similar, it was said, to that caused by the meteors of 1833. Apprehending the approach of Judgment Day, crowds besieged the shop of Mr. Petty, our preaching tailor, invoking his prayers. Methodism reaped a harvest from the comet."

On June 3, 1858, Italian astronomer Giovanni Battista Donati (1826–1873) discovered a comet in the constellation Leo. He called the comet faint and described its faint glow as 3 arcminutes across. Observations continued over the coming days and there was no indication, from the many notations of "very faint," that this object would become a spectacle. Become a spectacle it did, however, swinging past Earth at a distance of 80.5 million km on October 10 and remaining visible for months. Comet Donati (C/1858 L1) became one of the brightest comets of the 19th century, after the Great Comet of 1811.

Donati's approach to both Earth and the Sun caused it to brighten rapidly during July and August, visible in evening twilight. The brilliant apparition hung a tail of some 35° length over the horizon for observers in the north, who saw that it moved among the stars of the Big Dipper by September. Numerous observers recorded details of the comet, and a new era of comet science began on September 27, when English photographer William Usherwood made a 7-second exposure of Donati using an f/2.4 portrait lens. It was the first photo ever taken of a comet. On September 28, American astronomer George P. Bond (1825–1865) created the first-ever photo of a comet made through a telescope when he captured a 6-minute exposure of Donati with the 15-inch refractor at Harvard College Observatory.

In Jonesboro, Illinois, senatorial candidate Abraham Lincoln observed Donati's Comet from the porch of his hotel, the night before the Third Lincoln-Douglas Debate, which pitted him against the diminutive Stephen A. Douglas. The comet faded in October and was visible primarily in the Southern Hemisphere. Nonetheless, it had made a terrific impression on astronomers and the lay public alike.

Just three years later came along another Great Comet when Australian farmer and amateur astronomer John Tebbutt (1834–1916) spotted a "fuzzy star" on May 13, 1861, in the constellation Eridanus. The Great Comet of 1861 (C/1861 J1), also called Comet Tebbutt, was confirmed the next night when this nebulous patch moved against the stellar background.

The Great Comet of 1861 was seen as a portent of doom by some in the United States, given that the enormous Civil War in America had commenced with the firing on Fort Sumter, South Carolina, just a few weeks before the comet's discovery. The comet was destined to sweep through the constellations Orion, Taurus, Auriga, Lynx, Ursa Major, Draco, Boötes, Hercules, Lyra, Cygnus, and Cepheus. In June, the comet proceeded on its way to close encounter with Earth on June 30, at a distance of 19.8 million km.

Early in June, observers reported the comet's nucleus at about 2d or 3d magnitude. In Sydney, New South Wales, Australia, astronomer and clergyman William Scott (1825–1917) observed Tebbutt on June 8 and noted, "The comet is now sufficiently brilliant to be seen without a telescope 40 minutes before sunrise; its tail extends 18° in a direction 15° W. of South, one narrow stream of light extending twice as far as the rest of the tail. The nucleus is distinct and round, presenting no remarkable features."

At the time when the Great Comet of 1861 passed closest to Earth, June 30, numerous sky watchers made observations and recorded notes. Observing from Athens, Greece, German astronomer Johann F. J. Schmidt (1825–1884) noted the comet was "not as bright as Jupiter." But the English observer the Reverend Thomas W. Webb (1807–1885), famous for his celebrated *Celestial Objects for Common Telescopes* (1859), estimated its magnitude as being between those of Jupiter and Venus, which would have made it considerably brighter.

While observing the comet, Webb and his wife each noted a faint ray "of perfectly similar character to the tail, stretching under the square of Ursa Major, about 3° or 3.5° broad … and traceable about half way from the latter star to Arcturus: it pointed to the Comet, but in the twilight no connexion could be made out."

An Irish amateur astronomer in Dublin, J. M. Stothard, had just finished observing Jupiter when he

> was very much surprised at seeing a remarkable-looking nebulosity in the full glare of the twilight, in the northwest part of the heavens, which twilight was then so great as to render invisible all stars in those regions nearly as far as the zenith. I soon perceived the object a distinct

> and planetary-looking nucleus, which was well seen in an opera-glass, and shone out like the disk of a planet seen with low power with a dense halo round it.

As July commenced, brightness estimates of the Great Comet of 1861 ranged from comparable to that of Jupiter to closer to Saturn's. On July 1, astronomer James M. Gilliss (1811–1865) of the U.S. Naval Observatory peered through a bank of clouds, seeing a "pulsating light extending towards the zenith." Thinking the light was an aurora, he later discovered its true identity. In a better sky in England, an amateur astronomer with the surname Tidmarsh recorded that the comet showed "a huge tail [spanning] at least sixty or seventy degrees across the heavens. It could be distinctly traced across Polaris, and considerably beyond, nearly in the direction of Vega."

When Gilliss finally had a clear sky on July 3, he noted, "The constancy of the light near the nucleus was interrupted by flashings or pulsations, closely resembling those of the aurora." At Hamilton College Observatory in Clinton, New York, German-American astronomer Christian H. F. Peters (1813–1890) described a region near the comet's nucleus "filled with jets or flames, magnificently radiating from the nucleus, and bending before reaching the limb."

Estimates of the Great Comet's tail length by July 5 varied from 45° to 85° long. The comet was fading slightly, but still an incredible sight. Observers were still seeing great details in the comet's inner coma, near the nucleus, many commenting on one bright inner envelope of light and two fainter, outer envelopes. Amazingly, on July 6, Scottish explorer David Livingstone (1813–1873), in the fourth year of his Zambezi Expedition (in present-day Malawi), independently discovered the comet and described it as a "large comet in Ursa Major."

July 7, 1861, was a last hurrah for important observations of the comet. Traveling separately from Livingstone but in the same area, the explorer's physician John Kirk (1832–1922) spied the comet and wrote simply, "This night we got sight of a splendid comet in the Great Bear moving rapidly from the Sun." In Washington, Gilliss noted the tail at 25° long and recorded the luminous areas of the inner coma were "much smaller and fainter, and for the greater part of the time could scarcely be discerned at all as distinct from the general mass of light."

As the year wore on, the comet faded. By August, when the comet was in Boötes, the tail had shrunk to 2° and the magnitude of the nucleus was about 8. The big show was over. Observers mostly at professional institutions around the world continued watching Comet Tebbutt in spring of 1862, with the last observation on May 1. The Great Comet of 1861 was no more.

Twenty years passed and another Great Comet graced Earth's skies – a comet that would be one of the most spectacular of the 19th century. Secondhand accounts put the discovery of this comet at about September 1, 1882, by observers at the Gulf of

Guinea and the Cape of Good Hope. This brilliant object, whose earliest observations were chronicled by German astronomer Johann G. Galle (1812–1910) in 1894, would come to be known as the Great Comet of 1882 (C/1882 R1), or, more frequently, as the Great September Comet of 1882.

The comet shone brilliantly in the morning sky and was claimed by many to be as bright as Venus. On September 5, American astronomer Benjamin A. Gould (1824–1896) reported in the *Astronomische Nachrichten*, "Inquiry showed that it had been seen for several days by employees of the railroad and other persons whose duties required them to rise before daylight." First among astronomers to see the comet was Englishman William H. Finlay (1849–1924), who was observing an occultation of 5 Cancri by the Moon on September 8, in Cape Town, South Africa, when he independently found the comet. Finlay described it as a conspicuous object with a degree-long tail and a magnitude of about 3.

Tipped by his friend Finlay, American astronomer William L. Elkin (1855–1922), also in Cape Town, observed the comet on the morning of September 9. "It then appeared to the unassisted eye about as bright as a star of the third or fourth magnitude," he penned, "with a straight tail about 2.5° in length. The color of the comet's light struck me as remarkably white, perhaps contrasting it from recollection with comet Wells, which was of a brilliant golden hue."

Once astronomers calculated its orbit, they saw the Great September Comet would pass within 146 million km of Earth on September 17. When discovered, the comet was in the constellation Sextans; it would swing through Leo, Virgo, Hydra, Puppis, Canis Major, Lepus, and Monoceros over the coming 6 months.

Although the comet was moving into the morning sky, near the Sun's glare, after its discovery, observers still spotted it. The English chaplain Joseph Reed, on board HMS *Triumph* near the Cape Verde Islands, reported the crew's observations of the comet on September 12 and 13. "The comet was visible for only a few minutes before sunrise," he wrote. "The twilight prevented determining any length of the tail, but it appeared to extend through an arc of two or two and a half degrees. The whole of the coma is very brilliant, the nucleus surrounded by a still brighter ring; the tail was not curved."

The September Comet of 1882 continued to increase in brightness, so much so that it was visible in the daytime for more than 2 days. On September 17 John Tebbutt observed the comet when it was a mere 4° from the disk of the Sun, and it was "moving fast towards that luminary. The head and the tail for about a third of a degree were well seen." A short time later on the same day observer L. A. Eddie, in Grahamstown, South Africa, spotted the comet and was amazed. "So apparent was it to the naked eye, that one had but to look in the direction of the Sun when it could be immediately seen without any searching."

As the comet approached perihelion on September 17, it transited the face of the Sun for 1 hour 17 minutes. On the other side of perihelion, the comet emerged less than a degree west of the Sun's disk and became visible again. At Simons Bay, South Africa, Scottish astronomer David Gill (1843–1914) watched the comet after perihelion passage. "I was astonished at the brilliancy of the comet," he wrote, "as it rose behind the mountains on the eastern side of False Bay. The Sun rose a few minutes afterwards, but to my intense surprise the comet seemed in no way dimmed in brightness, but becoming instead whiter and sharper in form as it rose above the mists of the horizon."

Word of the comet's visibility reached the United States on September 19. At Princeton Observatory in New Jersey, astronomer Charles A. Young (1834–1908) attempted to use the 23-inch refractor to view the comet, but glare from the Sun prevented it. "The comet was beautifully seen in the 13-cm finder," he penned,

> which was screened from the glare by the tube of the great telescope. The nucleus was diffuse, not stellar, (magnifying power about 75); the first envelope was pretty bright and well defined, extending out on each side to form the tail, an the second envelope was easily visible, though rather faint. The interesting feature, however, was the pair of eccentric arcs connecting the two envelopes: they were not conspicuous, but I think there is no doubt as to their reality.

As the comet moved slowly away from the Sun, its tail lengthened. In Nashville, Tennessee, American astronomer Edward E. Barnard (1857–1923), chronicler of dark nebulae, noted a straight, long, 12° tail on September 22 and wrote, "A bright outline passes completely around the head, forming a parabolic curve of yellowish light which streaming backward traces the boundary of the tail."

The Great September Comet of 1882 began to show some curious behavior in October. Despite its fading overall brightness, the coma bristled with activity. Observers reported that the nucleus became egg-shaped, and on October 5 several sky watchers recorded a multiple nucleus, with three distinct stellar points. This caused many astronomers to watch the nucleus carefully throughout late October. English astronomer Ernest E. Markwick (1853–1925), observing with a 7-cm refractor in Pietermaritzburg, South Africa, noted on October 5, "South, preceding the comet's head, at this time were seen, about 1.5° distant, two wisps or pieces of nebulous-looking light." Several other observers monitored these unusual detachments of light, some describing the phenomenon as a "double comet."

The comet continued fading in November, although the tail remained some 20° long. Many telescopic observers reported seeing multiple nuclei in the comet, as many as six distinct points of light. The following weeks saw the comet continuing to slowly fade, to 5th magnitude by December and still a naked-eye object in January 1883. The following month, the Great Comet sank below naked-eye visibility and entered the realm of telescopic observers alone.

After the Great September Comet, 27 years would pass until one of the magical moments in cometary observing arrived. In 1910, two sensational comets would grace Earth's skies. The first was discovered, suddenly, by multiple observers who spotted it on January 12, 1910, as a bright object already visible to the unaided eye. The Great Comet of 1910 (C/1910 A1), also called the Great January Comet or the Daylight Comet of 1910, was off and running with a bang. The second comet was old, reliable Halley's Comet, visible again for the first time since 1835.

When diamond miners in the Transvaal of South Africa spotted the Great Comet of 1910 in predawn twilight, it was already a brilliant object, shining steadily at about magnitude −1. The comet did not become well known, however, until January 15, when Scottish–South African astronomer Robert T. A. Innes (1861–1933) received a telephone call from a Johannesburg newspaper informing him of other predawn observations.

By January 17 Innes glimpsed the comet when it was merely 4.5° away from the Sun. He called it a "snowy white object" with a tail 1° long. Daytime observers spotted the comet in Rome, Vienna, and Algiers. After perihelion on January 17, the comet moved sufficiently away from the Sun to become a brilliant evening sky target for Northern Hemisphere observers. Many who had seen this comet later mistakenly believed it to have been Halley's Comet.

On the evening of January 21, an English amateur astronomer, Ellison Hawkes (1889–1971), observed the comet and wrote:

> The picture presented in the western sky was one which will never be forgotten. A beautiful sunset had just taken place, and a long, low-lying strip of purple cloud stood out in bold relief against the glorious primrose of the sky behind. Away and to the right the horizon was topped by a perfectly cloudless sky of turquoise blue, which seemed to possess an unearthly light like that of the aurora borealis. High up in the southwest shone the planet Venus, resplendently brilliant, while below, and somewhat to the right, was the Great Comet itself, shining with a fiery golden light, its great tail stretching some seven or eight degrees above it. The tail was beautifully curved like a scimitar, and dwindled away into tenuity so that one could not see exactly where it ended. The nucleus was very bright, and seemed to vary. One minute it would be as bright as Mars in opposition, while at another it was estimated to be four times as bright. The tail, too, seemed to pulsate rapidly from the finest veil possible, to a sheaf of fiery mist.

These changes in the comet's appearance noted by Hawkes were presumably due to rapid changes in atmospheric seeing. The tail grew to observed lengths of 25° or more within a couple of days, and numerous sky watchers admired the comet the world over. Despite interference from strong moonlight as the month ended, the comet continued to put on a breathtaking display. Spectra made of the comet revealed the Great Comet of 1910 as a very dusty comet, and the strong curvature of the dust tail, as opposed to the relatively straight gas tail, gave rise to use of the term "glowing scimitar" to describe this comet in many observers' reports.

By mid-February the comet had faded to slightly brighter than the naked-eye limit, as it quickly receded from Earth and the Sun. By April the comet dimmed to magnitude 11, and in July the last observations of the Great Comet of 1910 took place when the icy chunk glowed feebly at magnitude 14.

But the Great Comet of 1910 was just the first of two rounds of excitement that year. Everyone knew that Comet Halley (1P/Halley) would return for the first time since 1835. The comet rose to naked-eye visibility about April 10 and would be a big deal in cometary science as it was the first apparition of Halley for which photographs and spectroscopic observations could be made.

The buildup to Halley's 1910 apparition was enormous. Of the 30 previous known visits of this comet to the inner solar system, none featured anything like the scientific buildup and the cultural hype of the 1910 visit (until 1985/6, of course). Astronomers began searching to recover Halley's Comet in the winter of 1908/9, this time using photographic plates and long exposures. On December 22, 1908, an astronomer at Yerkes Observatory, O. J. Lee, captured the comet on a plate, but the first to photograph and publicly announce the then-16th-magnitude comet's recovery was German astronomer Max Wolf (1863–1932), at the University of Heidelberg Observatory, on September 11, 1909.

The 1910 apparition was among the most observed visits of a comet in history. Observatories around the world imaged, made spectra of, and analyzed the comet throughout its entire appearance. Unlike some less favorable apparitions (including 1985/6), in 1910 the comet passed within just 22.4 million km of Earth, giving it an amazing brightness of magnitude 0 and a tail stretching nearly 100° across the sky at its greatest. The closest approach to Earth occurred on May 20.

The first visual observation of Halley in 1910 was by American astronomer Sherburne Wesley Burnham (1838–1921), who used the 40-inch refractor at Yerkes Observatory to spy the comet on September 15. The comet brightened slowly, reaching magnitude 12 by mid-November, 10 by mid-December, and 9 by late January 1910, as the Great Comet was stealing the show.

After passing behind the Sun, Comet Halley emerged in the morning sky in April and its brightness continued to rise. On April 12 Edward E. Barnard observed Halley at Yerkes and estimated its nuclear brightness at magnitude 8.3. Moving away from twilight and closer to Earth, Halley brightened in May, and the tail lengthened dramatically to 18° by May 3, 53° on May 14, 107° on May 17, and 120° on May 18 (Barnard's estimates).

On May 19, Halley's Comet transited the Sun. Although it was not possible to observe the comet against the Sun's disk, Halley must have brightened incredibly just before and after this event. Also on this day, Earth was expected to pass through the dust tail of Halley's Comet, an event that had sparked numerous irrational fears of death and disaster, of Earth's whole population smothered in cyanogen gas (which had been detected spectroscopically in the comet's tail). French astronomer Camille Flammarion (1842–1925) had famously remarked that this gas "would impregnate the atmosphere and possibly snuff out all life on the planet."

On the night of May 19/20, the comet's tail was visible in both the evening and morning skies. If Earth really did pass through the comet's tail, would a "supertail" glow spanning 360° be visible? Amazingly, a passenger on a ship in the Mediterranean Sea claimed to have seen a large, faint glow like the Gegenschein, some 45° high and 60° wide with a "pillar of light" at its center. Could this have been the glow of the comet's tail as Earth passed through it? Many years later, astronomer Zedeněk Sekanina calculated the comet's dust tail missed Earth by a Moon's-width but the gas tail indeed enveloped our planet.

Numerous phenomena that had been observed with Halley in 1835 and with many other comets visually were now captured on photographs with the 1910 apparition. Bright jets in and around the nucleus, tail detachments, variations in brightness from day to day, and other features kept astronomers busy with their findings. In April the comet passed close to the planet Venus, offering a convenient photo op. In June Halley's Comet dropped below naked-eye visibility. Astronomers continued to photograph and study it for a year, however; the last image was made at Lowell Observatory in Flagstaff, Arizona, on May 30, 1911, when the comet glowed weakly at magnitude 18.

The last Great Comet of the old days we'll visit is Comet Skjellerup-Maristany (C/1927 X1). Visible to the naked eye for 32 days, this brilliant comet was discovered on November 28, 1927. Evidence suggests that at least 10 people independently discovered the comet before that date, but none reported it to astronomical authorities, and therefore they squandered the opportunity to have their name go down in history. But the South African–Australian astronomer John F. Skjellerup (1875–1952, pronounced "shell-er-up") was awakened on November 28, went outside, decided to observe the sky, and spotted the comet. It turned out that his cat had knocked something over, creating the clamor. So this cat warrants listing as a major assist in the log of discovered comets.

On December 6 Argentine astronomer Eduardo Maristany found the comet, his observations aiding in its confirmation. So although the comet was for years known as the Great Comet of 1927 or Comet Skjellerup, it is now properly termed Comet Skjellerup-Maristany.

Figure 2.5. Comet 17P/Holmes moves like a floating blob across the field of the California Nebula (NGC 1499) in Perseus on March 8, 2008. The image was made with a 200 mm lens at f/2.8, a CCD camera, and stacked exposures. Credit: Gerald Rhemann.

Although the comet was bathed in bright twilight at this time, it shone brightly at about 2d magnitude and had a tail stretching 3°. The comet brightened rapidly, reaching 1st magnitude just a day later. This led to numerous reports by random people outdoors at the time who spotted the comet and reported it to all manner of authorities. A woman hiking the Sierra Madres in California spotted it when the Sun was hidden by a high peak. A timber industry worker near Flagstaff, Arizona, suddenly saw the comet and reported it to Lowell Observatory. The map of European observers lit up with viewings of the comet by the dozens.

The comet peaked in brightness sharply in mid-December and then began to fade. Many sky watchers saw it in broad daylight. Skjellerup himself observed the comet in the daytime, just 2° from the Sun, on December 15. He used a pair of binoculars and a small telescope, using a nearby chimney to block the dangerous rays from the Sun itself.

Indian astronomer P. R. Chidambara Ayyar observed the comet and described the head as a "nebulous mass with an extremely bright nucleus from which emanated two luminous, curved arms." He further declaimed, "I thought it was a huge giantess who had let loose the terrific glory of copper-colored mass of tresses, and who was running away with her back turned away from us." He and other astronomers judged it to be brighter than Venus, which lay nearby.

These incredible observations of Skjellerup-Maristany are centered on December 15, 1927. Recent analyses by amateur astronomer Joseph Marcus have attacked the issue of forward scattering of light from dust in comets as they pass between Earth and the Sun. This phenomenon happens with comets that are "backlit" by the Sun, which brighten significantly because dust and ice crystals in the comet significantly enhance the brightness of light passing through them or reflecting off them. This seems to have happened with this comet as well as several others including the Great Comet of 1910 and Comet West in 1975/6.

Skjellerup-Maristany was not particularly well placed; that is why it wasn't observed for very long. The last naked-eye sightings took place on January 3, 1928. By April 28 astronomers at Johannesburg made the last observations of the comet, which then glowed feebly at 14th magnitude.

Great Comets spanned an amazing reign in human history, from the time of Caesar to the emergence of astrophysics. (We'll explore the bright and unusual comets of the modern era in Chapter 4.) That period revolutionized astronomy, pushing it from a science of cataloging - understanding the simple differences between "specimens" - to real interpretation.

The late 19th century experienced an explosion of astrophysics, of photography, spectroscopy, and other tools allowing astronomers to start explaining the physical nature of objects in the universe. And this revolution in understanding was nowhere more meaningful than with comets. Seeing comets as physical bodies and gaining insight into their significance led astronomers to an adventure that would take them back to the very beginnings of the solar system.

3

What Are Comets?

The universe is a dynamic, active, often violent place. Stars, planets, and galaxies undergo rapid change, constantly worked on by their environments. Seeing Earth, the Sun, or the Milky Way as it was long ago is just not possible. Comets, however, give astronomers a precious storehouse of ancient information. They are relatively pristine, ancient blocks of ice that allow us to peer back in time.

Astronomers' knowledge of comets is still developing, but the knowledge base is vastly larger than it was a generation ago. Strangely, planetary scientists have gleaned much of what they know about comets without knowing precise numbers on some of their most basic physical properties. All we know about the shapes, sizes, and albedos (how reflective of sunlight the comet is) of the nuclei of comets has been gained from the small number of close visits to comets by spacecraft. In this way, only about two dozen comets are well known.

When astronomers observe comets from Earth, they face challenges like making assumptions about the object's albedo, which they can then use to calculate a presumed size. Recent measurements of a small number of comets such as 9P/Tempel 1 with the Spitzer Space Telescope, in 2005, suggested a low albedo of about 4 percent for that comet. The *Deep Impact* mission measured the comet at 7.6 by 4.9 km, and the inferred mass for Tempel 1 is about 75 trillion kg, on the basis of a density of 0.62 g/cm^3 calculated from data from *Deep Impact*, the spacecraft that encountered the comet. (By comparison, water has a density of 1 g/cm^3.)

But astronomers don't know the basics – mass, density, reflectivity – for nearly all comets. The best some have done is cleverly propose densities for some comets that are based on their nongravitational acceleration as they move about in their orbits. That is, jets can push them differently than they would move by the Sun's gravity alone, and observing this phenomenon allows astronomers to propose certain densities for them. In this way, astronomers have estimated the densities for comets

19P/Borrelly and 81P/Wild 2. A major rendezvous mission that will presumably be able to nail down all the parameters of its host comet is that of the European spacecraft *Rosetta*, en route to 67P/Churyumov-Gerasimenko, arriving in 2014.

Much of what astronomers know about the interior structure of a comet's nucleus was gained by the 2005 *Deep Impact* mission, the probe that unleashed an impactor into Comet 9P/Tempel 1. Planetary scientists were surprised at the amount of dust, as opposed to ice, released in the collision. The material was fine like talcum powder and contained clays, carbonates, sodium, and silicates, all detected spectroscopically. Astronomers determined the comet consists of about 75 percent empty space, suggesting the composition of a dirty snowbank. The presence of clays and carbonates suggested the existence of liquid water within the comet, rather than simply water ice, a surprising finding.

Other information came from the amazing 1994 impact of Comet Shoemaker-Levy 9 (D/1993 F2) into Jupiter. (We'll examine this momentous drama in detail later on.) From that unique event, astronomers found that on scales of about a kilometer, comets are held together relatively weakly, almost like glorified rubble piles, rather than incredibly solid, tough, hard-frozen bodies.

With comets, nothing comes easy. The chemical makeup of the nuclei also hides in a blanket of suspicion, with slim findings overall. *Deep Impact*'s spectrometer detected a range of compounds in Tempel 1, including silicates, carbonates, clay minerals in the smectite group, sulfides, amorphous carbon, and complex organic molecules called polycyclic aromatic hydrocarbons (PAHs).

But planetary scientists still know relatively little about the composition of cometary nuclei. They distinguish between volatiles (gases) and refractories (solids), and because the spectra of cometary nuclei are essentially featureless, astronomers mostly limit their spectroscopy to the coma. Astronomers do know the surfaces of comets are extremely dark, in part because some particles of larger size move out in front of others and cast shadows back onto the surface of the comet. It also results from the existence of extremely dark, carbon-rich material on the surface.

Recent studies of the Stardust material from Comet 81P/Wild 2 are intriguing. Some crystalline grains from the comet's tail formed at relatively high temperatures, suggesting that some material from the inner solar system mixed with ices in the outer solar system before comets formed. The similarity of some cometary dust to asteroidal material has mixed the origins of comets and asteroids together to some extent. (More on this later too.)

Planetary scientists do know quite a bit about volatiles in the coma – that's the part that's easiest to study. The race to discover what gases made up the coma began with the pioneer Fred Whipple himself, who proposed that escaping gases from a comet's nucleus produce forces on it and push it toward perihelion slightly earlier than would otherwise be expected. Whipple's model of ices subliming into gas and producing the large, visible coma and tail of a comet described water ice as dominant, and others – ammonia, methane, and others – also present. Whipple and his colleagues also proposed that ices locked up in the comet's nucleus had originated in interstellar space and so held special clues about the origin of the solar system.

Spacecraft armed with mass spectrometers flying alongside Comets 1P/Halley, 21P/Giacobini-Zinner, and others have provided good data, adding to a battery of ground-based measurements from the radio and optical parts of the electromagnetic spectrum.

The survey is young, as relatively few comets have been studied. But astronomers know that comets contain more than 80 elements or compounds, from simplest hydrogen to complex amino acids, the building blocks of proteins that in turn build self-replicating molecules – life. Our sampling of comets thus far turns up a garden of chemical compounds, including water ice, carbon monoxide, carbon dioxide, methane, acetylene, ethane, methanol, formaldehyde, formic acid, methyl formate, acetaldehyde, ammonia, hydrogen cyanide, isocyanic acid, hydrogen isocyanide, acetonitrile, hydrogen sulfide, carbonyl sulfide, sulfur dioxide, and sulfur. All but one of the compounds found in comets (sulfur) are also found in the interstellar medium, the thin soup of space surrounding our solar system and lying between it and the rest of the stars in our galaxy.

But the story of which compounds exist in comets is complicated by the fact that many of these are fragments of still larger molecules that must have existed in the comet's nucleus. So planetary scientists are in the early days of sifting through the meaning of the chemistry they find in comets. To complicate matters even more, observations of Comet Hale-Bopp (C/1995 O1) demonstrated that the observed chemical species varied with the comet's distance from the Sun.

When it comes to refractories – dust – the story is even murkier. (Ahem.) Observations have revealed cometary dust includes crystallized olivine group minerals along with pyroxene group species. Spacecraft observations of Comet 1P/Halley revealed grains of so-called CHON particles, a mixture of carbon-hydrogen-oxygen-nitrogen-rich material.

Unfortunately, getting a grip on cometary particles is difficult. Many professional collections of micrometeorites probably contain samples of cometary dust. Gutters on your house probably contain a very small quantity of them. But the proof that such tiny meteoric particles are connected to comets is tenuous at best. Dust grains in micrometeorite collections appear to be cometary in nature mostly because planetary scientists tend to think they look like what tiny pieces of comet dust ought to look like.

Progress occurred in 2006 when the *Stardust* mission returned samples of cometary dust locked in aerogel from its encounter with Comet P/81 Wild 2 and dropped the canister unceremoniously into the Utah desert. Some 150 scientists examined the samples and found, among other things, that Wild 2's dust contained more long-chain hydrocarbons than have been detected in the interstellar medium. They found small numbers of CHON particles, and no silicate hydrate minerals nor carbonates. They found abundant silicates such as olivine group minerals, anorthite, and diopside.

Minerals like olivine, anorthite, and diopside require high temperatures to form. So at first planetary scientists were baffled as to why they could exist as grains in comets, which after all spend their whole lives in a frozen state. Typical temperatures in the outer disk, where comets live, are far below the 1,000 K required to melt silicates and form these grains. So how did they get into comets? Astronomers believe they may have formed closer to the Sun and been blown outward by the solar wind, formed in the outer solar system by shocks or lightning, or major collisions that could heat and recrystallize material. To support the first notion, 2008 research revealed an oxygen isotope signature that suggests the tiny crystals in the comet formed in the inner solar system.

A 2011 study of Comet Wild 2 by University of Arizona astronomers uncovered iron and copper sulfate minerals in the comet – substances that require liquid water to form. So the notion that comets never warm enough to melt water ice must be untrue, at least for Comet Wild 2, and either a collision or radiogenic heating from the decay of radioactive elements must have heated portions of the comet enough to form these sulfates.

Astronomers' understanding of the composition of comets is in its relative infancy. Their notions of how comets formed and have moved throughout the solar system are a little more solid, based on hundreds of years of studying orbits and on more recent dynamical simulations.

In a basic sense, comets formed beyond the orbit of Jupiter and were pushed out into the massive shell known as the Oort Cloud or – for some comets inside the orbit of Neptune – were captured into closer orbits by the giant planets. This led to the Jupiter family of comets. But beyond these basic facts, astronomers know few details. Planetary scientists do not know, for example, the concentrations of comets in the Oort Cloud at various distances from the Sun. They also don't know the details of how distant comets in the Oort Cloud are knocked inward to make a pass close to the Sun.

The story of how comets evolve over time leaves much to be understood. Although astronomers know that some comets are highly evolved – that is, they've changed greatly from the early days – they don't yet have enough information to piece together the story of how comets evolve. (Astronomers agree that the comet with the shortest period of any, 2P/Encke, which orbits the Sun every 3.3 years, is a highly evolved comet.)

One stumbling block is that astronomers find that comets with very different evolutionary histories don't appear to be very different from each other. Comets that have clearly evolved differently and "lived" in widely different orbits nonetheless seem to have quite similar chemical compositions, the ratio of gas to dust in comets varies widely and is not correlated to environment, and the nuclei of all but the Jupiter-family comets have not been very well studied.

Planetary scientists often use statistical arguments to interpret the meaning of their observations. With comets, such arguments show that short-period comets must become inactive at some point in their lives, their volatiles used up by repeated passages close to the Sun. They also suggest that some percentage of "fresh" comets from the Oort Cloud must disintegrate as they make their first pass close to the Sun. These fresh comets certainly behave differently than veteran comets do, probably from the initial loss of the comet's outer layer, which until it warmed for the first time had been continuously subject to radiation for 4.5 billion years.

Understanding how the outer layers of fresh and not-so-fresh comets behave depends on computer modeling. A variety of factors influence how comets react to their first tastes of the Sun's warmth. The nature of a comet's orbit, the mixture of ices and dust, the processes acting on comets such as nongravitational jets, and so on – these all influence how the comet warms and grows its coma and tail. Many factors control how a comet's nucleus acts as it warms – the density and porosity of the material, the arrangement of ices and dust grains or pockets, the solidity of the ice or looseness of a rubble pile comet, and more. Astronomers don't know a great deal about how dust grains come off the surface of a comet, aside from realizing the forces acting on dust grains on the comet's surface and on those buried within ice are different. The dust's cohesiveness, its ability to stick together, would also influence how it behaves when liberated from the nucleus.

Some of the most interesting observations of comets have come from those that have broken up. Comets disintegrate from the Sun's gravitational pull, gravitational action by Jupiter or another large planet; from explosive events within the nucleus; or perhaps for yet-to-be-discovered reasons. The most celebrated breakup of a comet was the 1994 disintegration of Comet Shoemaker-Levy 9 (D/1993 F2), which sent 21 fragments slamming into Jupiter in July 1994.

But many other comets have been cast asunder into pieces. In 1995, 73P/Comet Schwassmann-Wachmann 3 broke apart into 4 large pieces. By March 2006 astronomers had observed 8 fragments; in the end, this comet fragmented into 66 separate pieces, most of which passed Earth in May 2006 at the scant distance of 11.9 million km.

In 1996 one of the most recent Great Comets, Hyakutake (C/1996 B2), which we will examine in detail in the next chapter, released a slew of small-sized particles. The comet's nucleus spans 2 km, but radio observations at the Arecibo Observatory

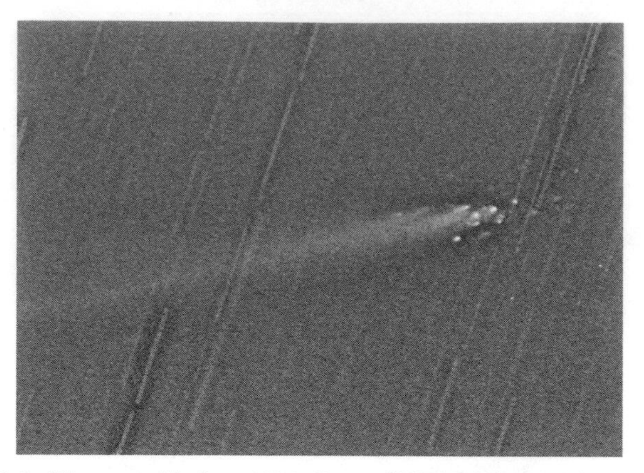

Figure 3.1. Discovered in late 1999, Comet LINEAR (C/1999 S4) spectacularly disintegrated close to perihelion on July 26, 2000; this image showing the nucleus in fragments was made on August 6, 2000, with the Very Large Telescope ANTU instrument, which revealed more than a dozen "minicomets." Credit: ESO.

in Puerto Rico showed repeated bursts of pebble-sized debris, and the observations of these small fragments allowed scientists to calculate a rotation period for the nucleus of 6.23 hours.

Many long-period comets and especially sungrazing comets have met their end by breaking up as they passed very close to the Sun. The joint European Space Agency-NASA spacecraft *SOHO* (short for *Solar and Heliospheric Observatory*) has captured images of many sungrazers ingloriously plunging straight into the Sun. Because it blocks the Sun's disk and allows observations very close to it, *SOHO* has become an unparalleled mechanism for discovering comets. Since it commenced operations in 1996, *SOHO* has collected data that have allowed researchers to discover more than 2,400 comets in the last decade and a half, making *SOHO* responsible for more than half the comets now known.

A spectacular example of a long-period comet that met its end observed by scientists comes from the story of Comet LINEAR (C/1999 S4) (Figure 3.1). Discovered in September 1999 by the U.S. Air Force/NASA/Massachusetts Institute of Technology project LINEAR (Lincoln Near-Earth Asteroid Research), the comet was headed for a closest approach to Earth on July 22, 2000, at a distance of 55.7 million km. It was on schedule for a perihelion passage 4 days later at a distance of 114.4 million km.

But the tiny comet had a nucleus spanning only about 0.9 km and had been outgassing volatiles and dust at a rate of about 1 cm per day. On July 5, 2000, Comet LINEAR (C/1999 S4) underwent a major fragmentation event, and then 15 days later it entirely broke apart. This comet became the first to disintegrate completely in front of the cameras and research scientists of the world.

Figure 3.2. A Hubble Space Telescope image of the fragments of Comet Shoemaker-Levy 9 (D/1993 F2) taken May 17, 1994, reveals the comet's train of 21 fragments that were poised to strike Jupiter. Credit: NASA, ESA, and H. Weaver and E. Smith (STScI).

Although from time to time comets break up, astronomers haven't yet seen enough of these disintegrations to draw many conclusions about them. They believe there's a small percentage chance that any comet could break up during a passage into the inner solar system. But do differences between groups of comets exist? Do more dynamically "fresh" comets break up the first time they swing inward close to the Sun? The exact mechanisms that control whether and how comets break apart are not yet well understood. Only those obvious breakups, like the gravitational twist of Jupiter causing Shoemaker-Levy 9 (Figure 3.2) to fall apart, are clear.

Astronomers point out a major contradiction from studying comets. On one hand, they are witnesses to conditions in the early solar system – a window into the distant and mysterious past. But they have yet to find out enough about how comets evolve over time to "subtract" their evolution and look at them in a purely pristine way, giving them that confident look back at how the solar system formed.

The real issues that could make comets incredibly valuable to understanding the early solar system are varied. For one thing, planetary scientists would love to know whether the ices that exist in comets are really pure – unaltered from the solar system's earliest days. In the early solar system, ices existed before they moved into comets. The Sun swelled, pulling in mass until it reached a flash point and commenced nuclear fusion, turning on as a star. Somewhere in this process astronomers believe a so-called accretion shock occurred, sending a wave outward, away from the Sun. Did the ices in the early solar system survive this accretion shock and migrate directly into comets? If so, they are literally material dating back to the solar system's beginnings. If they didn't – if they were changed significantly and newer ices formed in comets – then they can't show us conditions in the proto solar system.

Another question arises from understanding how important chemical reactions were in the part of the solar system's accretion disk where comets formed, when they were forming. If chemical reactions modified either the ices or the dust in comets as the solar system was gravitationally joining together as a disk, dramatic changes in the material that went into comets might have taken place.

Last, astronomers want to know what the relative abundances of the kinds of ices in comets can tell us about conditions in the early solar system. For example, scientists have created a wide range of ices in the lab and discovered that if ices form mainly by condensation, then temperatures play a big role in how much of different kinds of ices form. So this suggests that the kinds of ices that would form would be greatly influenced by how close or distant a protocomet is from the Sun. But what if ices survived the accretion disk shock portion of the solar system's formation and went directly into comets? Then they would not be forming ices directly from condensation of gas, and the question becomes irrelevant.

And to make a clear understanding even more difficult to achieve, planetary scientists believe protocomets from different areas of the solar system, in the early days of the solar system, probably intermixed. Trying to run the evolution of this idea backward to decipher what happened with cometary ices 4.5 billion years ago presents a horrifically complex, sticky problem – like trying to study and understand the formative DNA and genes of our long-dead ancestors.

More large, and potentially Earth-shaking, ideas about comets and their significance to us still loom mostly in the future. How much water did comets deliver to early Earth and contribute to the world's supply of oceans? We know that amino acids exist in comets and asteroids. Did comets play a role very early on by depositing relatively complex organic molecules that were precursors to life forming on Earth? Enormous amounts of research need to be done in these areas – and you should be careful when reading the latest single-study press release sent by some hungry institution that wants to claim to have singlehandedly solved the issue by the wave of a hand.

Significant problems exist with the idea of comets' contributing significant water to early Earth. One way planetary scientists have investigated this issue is to look at the ratio of deuterium to hydrogen (D/H ratio) in comets versus the same ratio in Earth's oceanic water. Deuterium, one of two stable isotopes of hydrogen (and often called heavy hydrogen), provides a good investigative lead. Today's Oort Cloud comets (for which astronomers have determined these ratios) have a very different D/H ratio than seawater on our planet.

While early Jupiter-family comets may have had very different D/H ratios and scattered inward and bombarded Earth, that hypothesis may be a long shot. Some scientists have suggested that asteroids rich in minerals containing hydrates could have bombarded the planet. But again, the volume of water Earth has makes that something of a head-scratcher.

How much of a hazard comets pose to citizens of Earth today is also something of an open question. Over the last 30 years planetary scientists have become increasingly aware of the large numbers of Near-Earth Objects (NEOs) that pose impact dangers to our planet. The night I am writing this in February 2013, a small asteroid (2012 DA_{14}) is passing just 27,700 km from Earth's surface – closer to the surface than geosynchronous satellites – a record close approach for its size. That space rock is just 50 m across, half the size of a football field.

NEOs, asteroids or comets that could collide with Earth, sum to about 10,000 objects, and most of these are quite small. Undoubtedly many very small NEOs are as yet undetected. But the process of discovering and tracking NEOs has improved dramatically over the past two decades. At one point you could say with a straight face that the number of people who were employed full-time in this business was lower than the day shift at a typical McDonald's restaurant. Now search programs and monies have poured in from a variety of sources to support NEO research.

The small asteroid that just whizzed past Earth, 2012 DA_{14}, would have caused a regional disaster. But let's step back and consider Earth's overall track record with solar system impacts. If you go out and look at the Moon, you can see a pretty nice record of solar system impacts. The Moon has no erosional or other resurfacing mechanisms like Earth's, so we are able to see its battered face dating back a very long time into the early solar system. But Earth has no magic shield. It has been battered, too, many violent times, but because of erosion, plate tectonics, mountain building, water, wind, and other processes, we do not see more signs of our own bludgeoned past.

Case in point: the Cretaceous-Paleogene Extinction Event, which happened about 66 million years ago and is well known even to schoolchildren because scientists believe it killed off, among other things, the whole population of dinosaurs. In one intense, worldwide mass extinction, approximately 75 percent of species on the planet disappeared. It is now often abbreviated as the K-Pg Event (for the German Kreide, Cretaceous, and the abbreviation for Paleogene), or more familiarly as the K-T Event, the "T" representing the now outmoded term "Tertiary."

In any case, in 1980 a team of scientists led by American Nobel Prize winning physicist Luis Alvarez (1911–1988) and including his son, the planetary scientist Walter Alvarez (1940–), discovered a worldwide layer of iridium far greater than normal concentrations at the time of the K-Pg Extinction Event. Iridium is a rare, silvery white metallic element in the platinum family, which is abundant in asteroids and comets despite its great rarity on Earth. The Alvarez team proposed that a massive impact deposited iridium in the so-called K-Pg boundary layer in sedimentary rock worldwide, and that the impact produced such terrible aftereffects that it was responsible for the mass extinction.

Viewed with skepticism by many scientists at first, the Alvarez hypothesis received regular gifts of supporting evidence. Geologists found pressure-shocked quartz and other minerals along the K-Pg sedimentary layer that suggested an enormous blast. Evidence arose for ancient, giant tsunami beds along the Gulf Coast and the Caribbean Sea. The K-Pg boundary layer in rock became more pronounced in the area of the southern United States, suggesting that an impact may have occurred in that area.

The enormous breakthrough arrived in 1990 when geologists confirmed the existence of the Chicxulub Crater, its scars now residing underground and underwater, centered near the town of Chicxulub on the Yucatán Peninsula of Mexico. The ghostly scar of the oval crater, spanning an average of 180 km constitutes the smoking gun of the K-Pg impact. It coincides perfectly with the destruction of the dinosaurs and many other species, and the depositing of the iridium in the K-Pg boundary.

American geologist Paul Renne (1957–) analyzed the data on Chicxulub exhaustively and found an impact date of 66,038,000 million years ago, plus or minus 11,000 years, and believes the mass extinction occurred within 33,000 years of the impact.

The K-Pg impact was no ordinary impact. Every day material from comets and asteroids enters Earth's atmosphere in a continual rain that amounts to some 400 tons per day. Most of this material is small. The stones that produce bright meteors – "shooting stars" – flashing across the sky are typically sand-grain or pea-sized. Rarely, pieces of small solar system bodies the size of basketballs, cars, or even houses fall into Earth's gravitational embrace, and most of these typically hit the ground as meteorites, some spectacularly breaking apart into myriad stones or metallic pieces, littering a debris field with prizes sought after and found by lucky collectors.

But the K-Pg impact was no ordinary impact. In 2010 a team of 41 scientists spread around the globe revisited the impact data and together concluded that an asteroid strike definitely caused the extinctions and consisted of an impactor about 10 to 15 km in diameter, delivering a blow equal to 100 teratons of TNT, equal to a billion times the energy released by the atomic bombs dropped on Hiroshima and Nagasaki combined.

The dangerous effect didn't result from the physical punch in the face. (OK, I know – it certainly wouldn't have been good to be in what is now the Yucatán on that day 66 million years ago.) The really disastrous effects were due to a variety of events that happened afterward. The worst part followed the impact itself when a "nuclear winter" scenario followed an incredible amount of dust thrown up into Earth's atmosphere that blotted out the Sun and inhibited photosynthesis. In a decade-long period before the thick blanket of aerosols finally dissipated, numerous plants, phytoplankton, and animals relying on them as a food source would have disappeared.

Intensely hot ejecta thrown up by the blast would have reentered the atmosphere, possibly igniting oxygen as it fell to the ground and commenced intense, worldwide brush fires. This would have produced far more smoke, further clouding the atmosphere, and lots more carbon dioxide, causing a short-lived greenhouse effect period, warming the planet significantly.

Acid rain, enormous tsunamis, the fact that the asteroid hit a gypsum bed and would have produced clouds of sulfur dioxide aerosols, and other complications all contributed to the devastating effect. The bottom line is that through studying the K-Pg impact, planetary scientists now believe that about 10 km is an important threshold when it comes to NEOs. That's about the size that would, if it struck Earth, be more than a regional disaster – it would be a civilization killer.

Fortunately for us, in the wake of the K-Pg impact, some small, hardy mammals survived. Mammals about the size of rats, perhaps resembling the ground squirrels we see scurrying around our yards, made it. We owe our very existence to them.

So how many of those roughly 10,000 NEOs are large enough to wipe out civilization on Earth with a direct strike? We don't know precisely. The largest NEO is asteroid 1036 Ganymed, which has a diameter of about 32 km. But 848 asteroids with orbits carrying them close to Earth exist with a diameter of at least 1 km. Of these, 154 are classified as potentially hazardous asteroids that could at some future date make exceptionally close passages to Earth. And of the 93 comets on the NEO list, no one conclusively knows of one that has struck Earth. But it's very possible that on June 30, 1908, a cometary nucleus exploded over the lonely forests of Siberia.

What's come to be known as the Tunguska Event occurred along the Podkamennaya Tunguska River in central Siberia, in the present-day subject of Krasnoyarsk Krai, a region occupying 13 percent of Russia's territory. Suddenly, an enormous explosion thundered over the mostly uninhabited, forested landscape, flattening perhaps 80 million trees over 2,150 square km. An enormous flash and bang were seen and heard over a huge area of the country, stunning hundreds of thousands of people. The shock wave from the blast would have registered about 5 on the Richter scale.

Tunguska was first noticed about 7:17 A.M. that day when indigenous Evenks of the Russian North saw a blindingly bright column of bluish light sweeping across the sky. Some 10 minutes later a blinding flash of light and dull, heavy thudding akin to distant cannon fire followed. The sound rolled over the landscape in the direction of east to north. The air heated by the trail of the incoming object expanded and broke windows hundreds of kilometers distant. Fluctuations in atmospheric pressure were detected as far away as England. High-altitude ice particles floated, locked in suspension, and caused an eerie glow high in the night sky for several evenings after the event. Atmospheric transparency decreased for several months as a result of substantial dust kicked high into the atmosphere.

Eyewitness testimony was varied and recorded with enthralling details. The newspaper *Sibir* reported:

> We observed an unusual natural occurrence. In the north Karelinski village the peasants saw to the north west, rather high above the horizon, some strangely bright (impossible to look at) bluish-white heavenly

> body, which for 10 minutes moved downwards. The body appeared as a "pipe," i.e. a cylinder. The sky was cloudless, only a small dark was observed in the general direction of the bright body. It was hot and dry. As the body neared the ground (forest), the bright body seemed to smudge, and then turned into a billow of black smoke, and a loud knocking (not thunder) was heard, as if large stones were falling, or artillery was fired. All buildings shook. At the same time the cloud started emitting flames of uncertain shapes. All villagers were stricken with panic and took to the streets, women cried, thinking it was the end of the world.

The Tunguska Event, blast, or explosion, as it was variously known, largely remained a mystery – substantially because of the then-incredibly remote place where it occurred. Not until 13 years later did the Russian mineralogist Leonard Kulik (1883–1942) first visit the region. (Of course, delays due to events that took precedence – World War I, the Russian Revolution, and the Russian Civil War – also occurred.) When he arrived on scene in 1921, undertaking a survey for the Soviet Academy of Sciences, Kulik deduced the blast must have been caused by the impact of a giant object from space. Astonished by the enormity of the forest destruction and lured by the possibility of a large amount of recoverable iron from the offending meteorite, Kulik persuaded the academy to fund a proper expedition to investigate.

When his expedition returned in 1927, Kulik talked the local Evenks into taking him to the site of the central impact. The journey was long and difficult, and the locals would not pass all the way to the impact zone, fearful because of freshly created superstitions. Kulik pressed on and found, amazingly, that he could not find a crater at all. Instead, he discovered a huge zone of scorched, upright trees whose branches had burned off; this spanned some 8 km in diameter. (The surrounding trees for many kilometers in every direction were flattened like matchsticks.) More than 30 years later researchers determined the blast created a butterfly-shaped zone of devastation whose "wings" stretched across 70 km and whose "body" covered 55 km.

Over the next decade, Kulik led three more expeditions to the remote Siberian forest. He found holes and bogs he attributed to meteorite strikes but later discovered stumps in some of them, ruling out meteoritic craters. The first aerial images of the blast zone were taken in 1938. Later research trips uncovered tiny silicate and magnetite spherules in the soil. The high percentage of nickel found in some of these spheres strongly suggested a meteorite as the impactor. With no impact crater, however, it seemed clear that whatever whizzed in from space exploded in the air above the Siberian woods, raining down debris violently from above.

Later scientific analyses of bog areas in the region also uncovered evidence of an extraterrestrial impact. Layers of sediment in the bogs were analyzed and found to contain different amounts of various carbon, hydrogen, and nitrogen isotopes than the layers from before and after 1908. The layer corresponding to the year 1908 also contained much larger amounts of iridium than normal, by analogy with what would later be found in the K-Pg sedimentary rock.

So if the Tugunska Event was caused by an impact, what happened to the body that struck Earth? Researchers could never find any evidence of solid bodies, meteorites, recovered from the site. Clearly, a significant object came hurtling in from space, exploded violently, and left almost no trace of itself save for the widespread damage that it caused. From 1908 onward, scientists were left with quite an impact puzzle.

In 1930 British astronomer Francis J. W. Whipple (1876–1943) – unrelated to the American Fred Whipple – proposed that perhaps the Tunguska impactor was not an asteroid but a comet. In contrast to an iron-nickel or rocky meteorite, a predominantly icy comet may have largely vaporized as it slammed into the lower part of the atmosphere, most of it disappearing into gas. The fact that Europeans saw a brighter than normal sky for several nights after the impact might also support this idea, the icy particles, water vapor, and dust scattered by the decomposition of a comet lingering in the upper atmosphere before being dispersed.

Since then, the idea that Tunguska was a comet that exploded in an airburst has washed back and forth, generally finding support but also coming into question. In 1978 Slovak astronomer Ľubor Kresák (1927–1994) linked the Tunguska impactor to the well-known periodic comet 2P/Encke, the parent body of an annual meteor shower called the Beta Taurids, which peaks in intensity during the last days of June. Kresák pointed out that the Tunguska Event occurred during that shower's activity and that the orbital trajectory of the impactor would have matched a stray fragment from that meteor shower.

A 1983 paper by American astronomer Zdeněk Sekanina argued the Tunguska object could not have been a comet because a comet would have disintegrated long before it approached the ground. American astronomer Christopher Chyba suggested that a stony asteroid could enter the atmosphere at high velocity, encounter a force that flattened and pancaked the leading edge, and create an explosion that blew the object apart, releasing nearly all its energy and leaving no crater. And a team led by Italian physicist Giuseppe Longo found that resin trapped in the trees of the Tunguska impact region contained grains typical of stony asteroids and atypical of comets. They also identified Lake Cheko, a small body of water in the region, as a possible crater tied to the Tunguska Event, which may have been caused by a small fragment sent reeling downward by the Tunguska airburst.

The comet hypothesis received a boost in 2010 when Russian-American physicist Vladimir Alexeev studied the Suslov Crater, one of the boglike holes found by the Kulik expeditions, this one spanning 32 m. Alexeev's team concluded this crater was indeed formed by the impact of a body from above. The surface was covered by permafrost, but lower layers revealed disturbed ice and, at the lowest levels, icy debris that may have been from the impactor itself, supporting the comet hypothesis.

Planetary scientists still debate whether the Tunguska Event was caused by a cometary nucleus that shattered in an airburst or an asteroid that violently exploded and left little trace – and of course the lines between comets and asteroids are becoming increasingly blurred. What's clear is that a small body of the solar system some 100 m across fell rapidly toward the Siberian forest and, 5 to 10 km above the ground, exploded violently with the force of 1,000 Hiroshima bombs.

Amazingly, while I was writing this chapter, on February 15, 2013, an unprecedented meteorite fall occurred in Russia. Making it really strange, the object is apparently the largest to encounter Earth's atmosphere since the Tunguska Event 105 years earlier. What's even stranger is that it took place a day before the very close passage to Earth of a NEO, the 50-m-diameter asteroid 2012 DA_{14}, and that the two events were clearly not related because the orbital dynamics of the bodies did not match.

As we've seen, lots of material enter Earth's atmosphere every day. Most of it never reaches the ground. The Russian meteorite fall was significant in a scientific sense because it was a large object that exploded in an airburst, and in a humanitarian sense it was a tragedy because about 1,500 people were injured, not struck by meteorite fragments directly, but injured by shattering glass from buildings. As the meteorite zoomed into the atmosphere, over the southern Urals and exploding over Chelyabinsk Oblast', it created a superheated column of air (dramatically increasing the air pressure), followed by a shock wave that knocked out windows all over the region. Injuries from flying glass were in some cases severe.

This case of any injury on Earth from an impacting solar system body is almost unprecedented. Rumors abound over injuries from early meteorite falls – a Milanese friar, a German boy, a dog even supposedly hit and "vaporized" by the Nakhla Martian meteorite in 1911. The only meteorite known to have hit a person is Sylacauga, which fell in the Alabama town of that name on November 30, 1954. The meteorite struck the house of Ann Elizabeth Hodges (1923–1972), badly bruising her in the midsection after bouncing off a radio. But the Russian fall of 2013 is far beyond anything ever known before – hundreds of people injured by the effects of a small solar system body.

The Chelyabinsk fall alerts us again to the very real dangers posed by NEOs. The dramatic fireball was captured on many incredible videos because of the abundance of car dashboard cameras in the region. Workplace videos showed the shock wave striking and sending employees into temporary panics. A zinc factory was damaged significantly by the sonic wave. Overall, as many as 3,000 buildings spread across six cities may have been damaged.

And all this occurred because a meteoroid with a diameter presumably of only a few meters slammed into the atmosphere at a velocity of about 15 km/sec and produced an airburst, some 15 to 25 km above the ground, with the explosive force of 500 kilotons of TNT, 30 times more powerful than the Hiroshima bomb.

The hazards of small bodies flitting around the solar system are very real, and never had the dangers been more amazingly demonstrated than with the unique bombardment event provided in 1994 as Comet Shoemaker-Levy 9 (D/1993 F2) slammed into the planet Jupiter's cloud tops.

I've had the privilege of knowing Canadian amateur astronomer and author David H. Levy (1948–) since the late 1970s, when he began writing for my homemade publication for sky watchers, *Deep Sky Monthly*. David introduced me to many people in the field in those early days, among them the wonderful planetary scientists Eugene M. Shoemaker (1928–1997) and his wife, Carolyn Shoemaker (1929–). Carolyn is one of the leading discoverers of comets in history, having found 32.

As I spent time in the 1980s at meetings or on trips to Lowell Observatory in Arizona, I got to know the Shoemakers and they were among the kindest, most brilliant people I've encountered – amazing human beings. Tragically, Gene was killed in an auto accident while researching impact craters in Australia, and planetary science lost one of its best friends.

Gene Shoemaker was particularly noted for in essence inventing the science of impact geology. When Gene began his studies it wasn't even clear that Barringer Meteor Crater in Arizona was an impact crater. He found shocked quartz and demonstrated it was indeed an impact before serving as first director of the U.S. Geological Survey Astrogeology branch in Flagstaff, training astronauts for the *Apollo* missions, and much more.

In the early 1990s Gene and Carolyn were busily finding comets and asteroids with the 16-inch Schmidt telescope at Palomar Mountain Observatory in California. David Levy had initiated a program of spending time helping the Shoemakers at Palomar, shuttling back and forth between California and his house in Tucson, Arizona. On March 24, 1993, the Shoemakers and Levy found a new comet on a plate exposed with the telescope, and it became known as Shoemaker-Levy 9 (the team had previously discovered eight short-period and two long-period comets together).

Incredibly, the Shoemaker-Levy team heard stunning news once astronomers calculated an orbit for the newly found object. It was orbiting Jupiter. Never before had a comet been found that was circling a body other than the Sun. Calculations showed the comet was probably captured by Jupiter's gravity some 20 or 30 years before. Sometime during the 1960s or early 1970s the comet strayed within Jupiter's so-called Hill Sphere, the region of its gravitational influence in which it will secure a smaller body in orbit. (The term derives from American astronomer George W. Hill [1838–1914], who defined the concept.) The comet's orbit about Jupiter was a strange, large, flattened ellipse: S-L 9 orbited the planet about once every two years and reached a point at its most distant of 49 million km from the giant planet.

As they gazed at the discovery images of SL-9, astronomers knew they had a very unusual comet. SL-9 appeared to have multiple nuclei strung along an area measuring 50 by 10 arcseconds of sky, about 1/36 the diameter of the Full Moon. Because the comet was close to Jupiter in the sky, Brian Marsden suggested the possibility of Jupiter's gravitationally tearing the comet apart.

Only when astronomers studied the orbital dynamics of SL-9 carefully did they find that apparently the comet had passed extremely close to Jupiter on July 7, 1992 – moving just 40,000 km above the planet's cloud tops. Planetary scientists believe this close passage was the one that caused SL-9 to shred apart into fragments. Once they studied the orbit even more carefully, it became clear that in July 1994 the comet would skim much closer to the planet and potentially dive into the cloud layer in a spectacular series of impacts.

This was a first in the history of astronomy – seeing a comet smash into a planet. A whirlwind of activity arose as astronomers all across the globe began to study the comet carefully and to prepare for the big event. The Shoemakers and Levy appeared on magazine covers, on television, and in numerous radio programs to discuss the unprecedented cometary destruction. The unique opportunity to study Jupiter's upper atmosphere was not lost on planetary scientists, as the cometary nuclei would act as probes, plummeting into the atmosphere and causing disturbances far below the swirling cloud tops.

Astronomers believed the nucleus of SL-9 was about 5 km across, and when it broke apart into 21 recognizable pieces (identified by letters), they ranged from a few hundred meters to 2 km in diameter.

As the impacts approached, a battery of Earth- and space-bound telescopes trained their eyes on Comet Shoemaker-Levy 9. The Hubble Space Telescope, with optics that had been repaired from the original set; the ROSAT X-ray observatory; the *Galileo* spacecraft; the *Ulysses* satellite, and even distant *Voyager 2* prepared for observations.

With the world watching, Fragment A slammed into Jupiter's upper atmosphere on July 16, 1994, ripping into the clouds at 60 km/sec and causing a hot fireball of 24,000 K and an upward-rising plume that reached 3,000 km tall. Amateur astronomers were in on the party, too – the "scars" from SL-9's impacts were readily visible in ordinary backyard telescopes. I recall the thrill of seeing numerous impacts as the event unfolded from the comfort of my Wisconsin backyard, using an 8-inch Celestron Schmidt-Cassegrain scope and a variety of magnifications.

Spacecraft and high-resolution ground-based telescopes imaged the flashes of the impacts, whereas backyard astronomers saw the series of dark blobs on Jupiter's cloud tops that resulted from each strike. Over the following 6 days, a total of 21 fragments plunged into Jupiter and created a trail of dark spots across the planet as big Jupiter rotated. It was a thrilling and unique event in the observed history of the solar system. The largest fragment, G, struck the planet on July 18 with the force of 6 million megatons of TNT, 600 times the complete nuclear arsenal on Earth, and left a dark scar spanning 12,000 km – roughly the diameter of Earth.

Observations of the SL-9 impacts revealed a trove of science about comets and about Jupiter. As the fragments punched through the Jovian atmosphere, they churned up material from below. Spectra revealed the presence of diatomic sulfur (S_2) and carbon disulfide (CS_2) in the atmosphere of Jupiter, both new discoveries. Astronomers detected a range of other compounds, including ammonia (NH_3) and hydrogen sulfide (H_2S), and emission from plentiful iron, magnesium, and silicon, the latter three probably from the comet.

Energetic shock waves rippled across the planet's clouds for as long as 2 hours following each impact. Radio observations revealed a sharp increase in continuum emission at the 21-cm wavelength following each impact, signaling the presence of synchrotron radiation caused by electrons released by the impacts and spiraling into the Jovian magnetosphere at ultrahigh velocities. After the fragment known as K struck Jupiter, observers noted an aurora in the vicinity.

To their surprise, astronomers did not detect significant amounts of water following the impacts. They believed the fragments might plunge into a layer of water below the cloud tops in Jupiter's atmosphere, thrusting some upward to be detected in the resulting spectra. But little water showed up, suggesting the cometary fragments probably did not dive deeply into the Jovian atmosphere but exploded in airbursts higher than was expected.

For astronomers, amateur and professional alike, the wake of the SL-9 impacts lived on for months. The dark scars – Jupiter's "black eyes" – were visible long afterward and sometimes rivaled the planet's famous long-lived storm, the Great Red Spot, for ease of visibility. Enriched amounts of ammonia, carbon disulfide, and other compounds floated around the Jovian atmosphere for as long as 14 months before returning to their normal levels. The planet's stratospheric temperatures rose after the impacts, fell sharply afterward for about 3 weeks, and then returned to normal.

For many, the impact of Comet Shoemaker-Levy 9 into Jupiter was not just a unique and exciting event in solar system history. It was also a wakeup call to the real dangers of impacts occurring in the solar system and a reminder that Earth does not hold a card of immunity in this dangerous game.

But it also reminded us of the incalculable value of Jupiter to life on Earth. Often referred to as a "failed star," Jupiter is no such thing. It is vastly short of the mass needed to become a star. But as the giant planet of our solar system it does provide a sort of sweeping service when it comes to small bodies continuing to fly around and run into things. Jupiter's strong gravitational influence draws comets, small asteroids, and meteoroids in and prevents them from posing a threat to the inner solar system, including Earth. Without Jupiter cleaning the streets, as it were, it's certain that many more large impacts would have occurred on Earth than actually did, and we all might not be here now to write books or read them.

After SL-9, astronomers identified other comets that, at least for a time, have orbited Jupiter. The comets in this subset of the Jupiter-family group are called quasi-Hilda comets because they are associated with the Hilda family of asteroids, which have an orbital resonance with Jupiter. A dozen comets currently belong to the group, including 82P/Gehrels, 147P/Kushida-Muramatsu, and 111P/Helin-Roman-Crockett.

The orbits of these comets are not stable over long periods, so comets in this group will come and go over long intervals. Most of these objects are small, however: SL-9 was rare in part because it was large. Nonetheless, Jovian impacts are probably a steady part of the solar system's history. Numerical studies suggest that a comet of 0.3 km diameter hits Jupiter once every 500 years, and that a 1.6-km-diameter comet strikes the planet on average every 6,000 years. On July 19, 2009, another black spot appeared on Jupiter, and analyses suggest this time it was a small asteroid that struck the planet.

Astronomers have learned a great deal about the behavior of comets by watching some of them break up. Their most fruitful era of rapid discovery, however, was almost a decade before the appearance of SL-9, in the mid-1980s, when the most famous comet in history approached the Sun on one of its graceful 76-year orbits.

Cometary science faces a major challenge unlike other areas of astronomy. Observers can either try to deduce details about a very small, icy body that is very far away or observe it when it's relatively close to the Sun, is warmer and more active than usual, but that very activity obscures details of the nucleus they are trying to understand.

The best of all worlds is to send a spacecraft to analyze a comet close-up, and that's what happened for the first time as the world anticipated the return of Halley's Comet (Figure 3.3) in 1985 and 1986. We've already seen that six spacecraft encountered Halley late in 1985 and in 1986, beginning with NASA's redirection of the *International Cometary Explorer* (*ICE*) spacecraft and ending with the extensive imaging produced by the *Giotto* probe. We've also seen that as apparitions of Comet Halley go, the 1985/6 one was pretty unfavorable from the standpoint of orbital dynamics. At perihelion on February 9, 1986, the comet was placed nearly behind the Sun as seen from Earth.

Despite the challenges, the opportunity to deepen knowledge of comets was astounding with Comet Halley, and it coincided with massive public interest and support, and the technological ability to build and launch probes that would – astronomers hoped – answer some fundamental questions about comets.

Figure 3.3. In 1986, the European Spacecraft *Giotto* became the first probe to encounter a comet. On March 14 of that year the spacecraft flew within 596 km of Halley's potato-shaped nucleus, which measured 15 by 7 by 10 km. This image shows the dark nucleus (right) with the inner coma trailing to the left. Credit: Halley Multicolor Camera Team, Giotto Project, ESA.

Unusual challenges plagued mission designers. Halley, like most comets, has an orbit that is tipped outside the plane of the ecliptic, the thin disk in which the planets orbit the Sun. Sending a spacecraft outside this ecliptic plane requires an enormous amount of energy. Because of this, planetary scientists determined to encounter the comet at the point it intercepted the ecliptic plane, on March 13, 1986. The five spacecraft that encountered Halley in March 1986 from Europe, Japan, and the Soviet Union formed an unprecedented coordinated effort to study the comet in an international group, the likes of which planetary science had never seen before. NASA stunned and disappointed the science community by failing to line up a mission to Halley because of funding cutbacks and political squabbling. Instead, the Americans simply rerouted the existing *ICE* spacecraft to encounter Comet 21P/Giacobini-Zinner before it flew through Halley's tail, measuring particles in the process.

The Americans did step forward with leading an extensive ground-based campaign to observe Halley's Comet in conjunction with the spacecraft missions. Coordinated by NASA's Jet Propulsion Laboratory and led by astronomers Ray Newburn (1933–) and Jurgen Rahe (1940–1997), the so-called International Halley Watch focused on collecting and coordinating observations of all types – astrometry, nuclear imaging, wide-field imaging, visible light and ultraviolet spectrometry, infrared and radio observations, and more. And in a massive program, the effort coordinated observations made by amateur astronomers. The program worked splendidly well, and two dozen CD-ROMs containing the results were produced and distributed by 1993.

The time leading up to the modern apparition of Halley was filled with anticipation. Astronomers decided to mark another reasonably bright comet, 27P/Crommelin, as a test object for Halley observations. They refined their efforts with the new wave

Figure 3.4. Diminutive Comet 144P/Kushida provides a lesson in the difficulty of spotting comets in this image made on February 2, 2009, when it lurked in the center of the Hyades star cluster in Taurus. The comet's small, fuzzy coma is visible inside the V-shaped Hyades, right of the brightest star Aldebaran and above and right of the bright double star just below center. The imager used a 100 mm lens at f/5.6 and stacked exposures at ISO 800. Credit: Chris Schur.

of digital detectors, using a sophisticated chip called a charge-coupled device (CCD), readying for the big event. Astronomers knew Halley's orbit accurately enough to predict its position on the sky to within 1 arcminute – very accurately, and within the small fields of view of even very large telescopes. They also believed that Halley might brighten to be "recovered" visually by some time in 1982. So something of a circus – a contest – moved forward to see who could detect the comet first.

When astronomers failed to recover the comet during the first 3 months of 1982, a slight panic spread throughout the community. Scientists began to wonder whether Halley's nucleus might have disintegrated since the 1910 appearance. During the summer of 1982 the comet was unobservable because it lay too close to the Sun. But by autumn 1982 the search was on again, and a Caltech team announced their imaging of Halley's Comet with the 200-inch Hale Telescope at Palomar Mountain Observatory, equipped with a CCD camera, just 9 arcseconds from its calculated position: located 11 astronomical units from the Sun – some 1.6 billion km – and glowing feebly at just 24th magnitude. A month later a French team using the

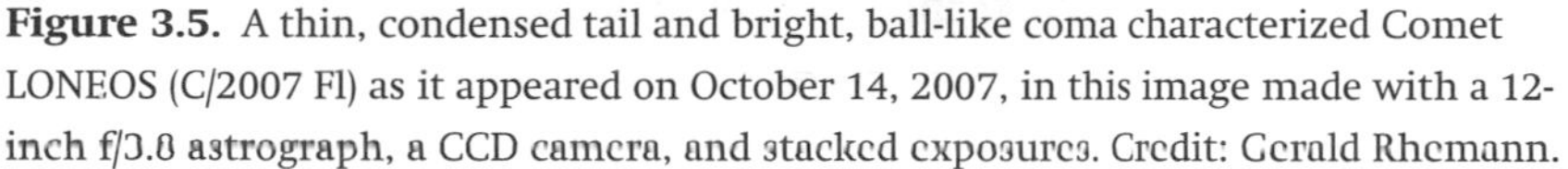

Figure 3.5. A thin, condensed tail and bright, ball-like coma characterized Comet LONEOS (C/2007 F1) as it appeared on October 14, 2007, in this image made with a 12-inch f/3.8 astrograph, a CCD camera, and stacked exposures. Credit: Gerald Rhemann.

Canada-France-Hawaii Telescope on Mauna Kea, Hawaii, imaged Halley. The modern era of observing Halley's Comet had begun.

Astronomers now watched the comet obsessively. In January 1983 astronomers at the European Southern Observatory in Chile observed a fivefold brightness increase in Halley. The comet had apparently undergone an outburst, blasting out molecules other than water, at that great distance from the Sun. Over the next few years a number of outbursts occurred as the comet slowly moved closer to the inner solar system. By 1985 not only professional astronomers were watching Halley's Comet, but amateurs too.

In the spring of 1985 a team from the University of Arizona captured the first spectrum of Halley, which revealed cyanide, and subsequent radio observations uncovered the hydroxyl radical. Planetary scientists could now focus on trying to understand the nature of the gases being emitted by Halley as it warmed and released ejections of material.

As astronomers planned their observations for late 1985 and early 1986, they understood that the time around perihelion in March would be essentially "blacked out" because of the comet's orbital geometry close to and behind the Sun, and that the preperihelion time would favor Northern Hemisphere observers, while Southern Hemisphere observers would have a somewhat better view after perihelion.

The first great window of scientific opportunity centered on September 1985. Innovative spectroscopic techniques using the 30-m IRAM receiver in Grenada, Spain, revealed hydrogen cyanide molecules in the comet. This marked the first-ever radio wavelength detection of a parent molecule released by a cometary nucleus. In December 1985 a team of Americans using the Kuiper Airborne Observatory detected water molecules in the comet, confirming that water is the principal constituent of Halley's Comet. Ground-based telescopes, meanwhile, focused on direct imaging and spectroscopy of Halley's nucleus.

Recall that the five spacecraft set to explore Halley were the Soviet *Vega 1* and *Vega 2* probes, the Japanese *Suisei* and *Sakigake* craft, and the European *Giotto*. The 1986 encounters confirmed the dirty snowball model of comets proposed by Fred Whipple and demonstrated the dominance of water ice in a comet's makeup. But they also threw a few surprises in the paths of unsuspecting planetary scientists.

Astronomers soon discovered that Halley's nucleus was exceptionally dark, irregularly shaped, and hotter than they anticipated. They had predicted a diameter of about 6 km for Halley's nucleus but measured, in situ, a larger, rectangular object spanning 8 by 7 by 15 km. Because the comet turned out to be larger than expected, scientists had to downgrade the assumed albedo of Halley from 10 or 20 percent reflectivity to about 4 percent. Up close, Comet Halley was almost coal-black.

The *Vega* and *Giotto* spacecraft carried mass spectrometers that analyzed cometary dust grains from Halley. Surprisingly, they found numerous primitive grains consisting of carbon, hydrogen, oxygen, and nitrogen atoms – the so-called CHON particles. These grains were the most abundant elements in the early solar system and offered a window back in time to the conditions of the solar nebula, the disk that formed our Sun and planets. Primitive hydrocarbons were abundant in these findings, and a *Vega* infrared spectrometer detected hydrocarbons in a gaseous or solid state.

These findings led to an amazing conclusion that tied in with the observed darkness of Halley's nucleus. Planetary scientists realized that the pure ices and dust making up the frozen block of a cometary nucleus were probably at least partially covered with a dark refractory layer produced by eons of solar radiation and cosmic rays acting on an organic crust. The icy snowball presaged by Fred Whipple turns out to be coated with dingy dirt from the interstellar medium.

Another surprising finding from the Halley spacecraft era was how complex and variegated the comet's nucleus was. Rather than a smooth, simple block, it proved to be highly active in some regions, unleashing violent jets of material, and inactive in others. The implications for temperature variations across the comet's surface are important, too. The active regions must be around the sublimation temperature for

Figure 3.6. Discovered in 2007 at the Mount Lemmon Survey in Arizona, Comet Boattini (C/2007 Wl) is a long-period comet that displayed an impressive coma. This image was made on July 2, 2008, using an 80 mm refractor and 12 stacked 3-minute exposures at ISO 800. Credit: Mike Salway.

ice, about 200 K, while the inactive regions must be in equilibrium with the temperature received by solar radiation, about 300 K at Earth's distance from the Sun.

The strange temperature findings, hotter than expected, implied something about the way comets may evolve. Perhaps, astronomers reasoned, the inactive portion of a comet increases over time - with more trips to the inner solar system - as ices continue to sublime away with these repeated solar encounters. This is one of the factors that lead to the ebbing away of the neat boundary between comets and asteroids, as now planetary scientists believe that some so called asteroids may be defunct comets.

But unanswered questions remain about the findings even from the Halley era. How is this dirty blanket of organic material that coats cometary nuclei, presumably from radiation, distributed over the nucleus? It is evenly spread? Does the organic coating hold up to repeated passages of a comet into the inner solar system? What is the coating's thickness? Does it reform quickly after the comet loses material by sublimation? Is the organic material less of a surface coating and more intermixed with ices deeper inside the nucleus?

Questions also lingered over another fundamental parameter, the rate at which the nucleus of Halley's Comet spins. Attempts in 1910 to measure the comet's rate of rotation were futile. Frustratingly, astronomers attempted to measure the comet's rotation in 1985 and 1986 with little success, other than that everyone agreed the rate was slower than 1 day. The spacecraft could not help because their encounter observations covered too brief a time. Ground-based observations came up with two divergent answers: 2 days or 7 days. Arguments erupted among the two camps. A compromise eventually came forth, suggesting a 2-day rotation with a slower, 7-day tumble superimposed on it.

Observations weren't limited to Halley's nucleus. The coma, consisting of a vast cloud of gas and dust, also received significant attention. Ground-based observations revealed hydrogen cyanide in the coma. *Vega*'s infrared spectrometer identified carbon dioxide. Ultraviolet and infrared spectrometry identified carbon monoxide, a common constituent of comets. Formaldehyde and possibly carbonyl sulfide were also detected.

Analyses of the coma also revealed, not surprisingly, abundant CHON particles. Whether these organics arose in the interstellar medium as complex molecules like polycyclic aromatic hydrocarbons (PAHs) or not can't be concluded. Such PAH molecules are considered pollutants on Earth but have been widely observed in the space between nearby stars in the galaxy. *Vega*'s spectrometer seems to have identified two PAHs in Comet Halley, naphthalene and phenanthrene. Strangely, however, these molecules were not observed in more recent bright comets such as Hale-Bopp (C/1995 O1) or Hyakutake (C/1996 B2).

Amazingly, the spacecraft observations of Halley's Comet also allowed astronomers to study the abundances of isotopes within the comet. These elemental variants with different numbers of neutrons provide the opportunity to gauge information about when the molecules containing various isotopes formed.

In Halley, astronomers found a relative abundance of deuterium ("heavy hydrogen"). This important discovery suggests the water ices trapped in a comet's nucleus were subjected to the same processes observed in molecules and ions in the interstellar medium, *before* being incorporated into the comet. Primitive processes in deep space put together the atoms, molecules, or even compounds that were later incorporated into the nuclei of comets, confirming that these icy denizens of the deep are indeed windows into the staggeringly distant, mysterious, deep past of the cosmos.

4

Comets of the Modern Era

The last Great Comet we explored was Comet Skjellerup-Maristany (C/1927 X1), which lit up Earth's skies during the final weeks of 1927. If we fast-forward 30 years, we come to the first Great Comet of the modern era, Comet Arend-Roland (C/1956 R1). Belgian astronomers Sylvain Arend (1902–1992) and Georges Roland (1922–1991) discovered the comet on photographic plates made on November 8, 1956. At the time of discovery, Arend-Roland already glowed at 10th magnitude and had a short tail. But the astronomers delayed announcing the discovery until November 19, and thereafter other images of the comet turned up, as early as September 11, 1956.

Comet Arend-Roland brightened slowly. Its orbit made clear that the comet would pass closest to Earth on April 21, 1957, at a distance of 85.3 million km. This followed its closest passage to the Sun on April 8. By year's end 1956 Arend-Roland was glowing at about 9th magnitude and sported a tail stretching just 8 arcminutes long. The first few weeks of 1957 saw the comet continue to brighten slowly, reaching magnitude 8.5 by late February. By February 27, the comet slipped too close to the Sun to observe, and astronomers had to wait until it reemerged in the morning sky.

In his masterpiece *Cosmos*, American astronomer Carl Sagan (1934–1996) wrote about a phone call he received while working as a graduate student at Yerkes Observatory in 1957. A persistent ringing pushed Sagan into answering the phone. A voice on the line, revealing an intoxicated caller, queried the young astronomer about some object in the sky. "Lemme talk to a shtrominer," the voice demanded. "Well, see, we're havin' this garden party out here in Wilmette, and there's something' in the sky. The funny part is, though, if you look straight at it, it goes away. But if you don't look at it, there it is." When Sagan informed him he was looking at Comet Arend-Roland, the caller asked what a comet was. When Sagan replied, "A snowball one mile across," the caller retorted, "Lemme talk to a *real* shtrominer!"

By April 2, 1957, astronomers recovered Arend-Roland and the comet shone at about 2d magnitude with a tail some 5° long. After perihelion on April 8, the comet continued to brighten as it drew closer to Earth. An image made at Mt. Stromlo Observatory in Australia on April 11 showed the comet's nucleus split in two, but visual observers did not seem to notice this event.

When Arend-Roland was closest to Earth, observers everywhere were gazing at the celestial visitor, which in late April dazzled at about 1st magnitude. From April 21 onward, the comet's brilliant tail stretched some 30° long, and, curiously, a bright sunward-pointing tail, called an antitail, also formed. Amazingly, the antitail itself stretched some 15° long. Logic explains how the solar wind pushes the gas and dust in normal cometary tails away from the Sun. So what could possibly cause a tail to point in the opposite direction – *toward* the Sun? Planetary scientists believe antitails consist of large dust particles that are minimally affected by the solar wind or the Sun's radiation pressure. They simply aren't pushed back away from the Sun. The fact that they remain suspended in the comet's orbital plane allows them to form a disk. As Earth passes through the comet's orbital plane, the disk is viewed edge-on and appears as a sunward-pointing "spike."

Arend-Roland's curious antitail remained visible until May 2. By that time the orbital geometry began to change and the antitail swung around the comet's nucleus and was no longer visible. The comet itself faded to the limit of naked-eye observers by May 18 and to a binocular object by June 3. Observatories continued tracking the retreating comet well into the following year. The last observation occurred on April 11, 1958, when the U.S. Naval Observatory astronomer Elizabeth Roemer (1929–) estimated the magnitude of the nucleus at 21.0. Curiously, the first program of the longest-running show about astronomy ever produced, the BBC's *The Sky at Night*, featured host Sir Patrick Moore (1923–2012) talking about Comet Arend-Roland, on April 24, 1957.

When two Japanese observers first glimpsed our next Great Comet, they had no idea it would become one of the most sensational of the modern age. Within 15 minutes of each other, on September 18, 1965, Kaoru Ikeya (1943–) and Tsutomu Seki (1930–) each glimpsed a fuzzy object that was telescopically faint. Ikeya was an amateur astronomer who worked at a piano factory; Seki was an astronomer who is now director of the Geisei Observatory in Kōchi, Japan. Thus was born what became Comet Ikeya-Seki (C/1965 S1). Note: This Great Comet should not be confused with another, fainter Comet Ikeya-Seki discovered by the pair later and designated C/1967 Y1.

Soon after this comet's discovery, astronomers studying the orbit realized Ikeya-Seki was a Kreutz Sungrazer. By the end of September excitement over the comet rose as scientists planned to observe it from high altitudes and with rocket launches. The comet would scoot past the Sun on October 21 at the minuscule distance of about 450,000 km – slightly greater than the distance between Earth and the Moon.

This suggested an immense brightness at its closest approach, and NASA administrators were so taken by the prospects they planned to alter the agenda of the planned *Gemini 6* mission to observe Ikeya-Seki. Unfortunately, the mission was delayed and in the end didn't coincide with the comet's visibility.

Ikeya-Seki attained naked-eye visibility on October 1, right on schedule, at about 6th magnitude. Two weeks later the comet had brightened to 2d magnitude with a 10°-tail and was an impressive sight. Spectacularly, Comet Ikeya-Seki brightened so rapidly that it became visible in broad daylight as it approached perihelion. Observers who held up a hand to block out the Sun's disk could see the wisp of the comet's tail nearby. When the comet was close to perihelion on October 21, the French astronomer Gérard de Vaucouleurs (1918–1995) observed the comet just 2° away from the Sun's disk and wrote that it displayed "a very bright nucleus with a silvery tail of one to two degree length." De Vaucouleurs described the comet as shining at about magnitude –10 at this time.

Soon thereafter, astronomers noticed that the comet's tail began to curve. And, curiously, Japanese astronomers using a coronagraph were able to observe Ikeya-Seki very close to its actual perihelion passage, a small distance away from the Sun's disk. They witnessed the comet's nucleus's breaking into three parts as the object neared the Sun.

After perihelion, during the last days of October, Ikeya-Seki faded but continued showing an impressive tail. Astronomers estimated the comet's tail at 45° on October 28 and 60° by the last day of the month. Ikeya-Seki began to fade substantially in November 1965. The comet was still visible to the naked eye on November 4, but the tail had shrunk to 20°. The American astronomer Howard Pohn of the U.S. Geological Survey discovered a secondary nucleus, and astronomers soon found a suspected third nuclear brightening. The secondary nucleus remained visible until January 1966. These observations reinforced what the Japanese viewers had witnessed close to perihelion.

By late November, Comet Ikeya-Seki had dropped below naked-eye visibility. On November 26 it glowed faintly at magnitude 7.5 and displayed a tail some 15° long. Southern Hemisphere observers continued to follow it for several more months.

Five years later, another Great Comet graced Earth's skies. Armed with a simple 5-inch Moonwatch telescope – an inexpensive instrument designed to follow satellites in the wake of *Sputnik* – the South African amateur astronomer John Caister Bennett (1914–1990) found a new comet on December 29, 1969, after observing for only 15 minutes. At this point the hobbyist Bennett, who had begun searching for comets in 1960, had spent more than 330 hours looking for comets without finding one. His luck had changed. The fuzzy blob he spotted in his telescope eyepiece was designated Comet Bennett (C/1969 Y1). This should not be confused with a later Comet Bennett, C/1974 V2.

Bennett was acutely aware of the lure of discovering a Kreutz Sungrazer and so was searching low in the sky, just 24° away from the South Celestial Pole, in the obscure constellation Tucana. Bennett followed the comet the next night and carefully noted its direction of travel, seeing that it wasn't a sungrazer. However, it was destined to become a very bright comet. Astronomers calculated Bennett's orbit and found it would reach perihelion on March 20, 1970, at a distance of some 80.5 million km. More importantly, the comet would pass 103.2 million km from Earth on March 26, at which time the magnitude would be about 0, as bright as most of the brightest stars in the sky.

By mid-January 1970 Comet Bennett was approaching naked-eye visibility. Throughout February the comet's brightness and size increased dramatically. On February 9 the magnitude was 5.5 and the tail a mere 1°, but by month's end Bennett had brightened to magnitude 3.6 and with a tail 2.2° long. The comet began to impress observers worldwide by mid-March 1970, when it grew to magnitude 0 and displayed a tail stretching 10°.

After perihelion passage, Comet Bennett remained stable in brightness as it moved away from the Sun but simultaneously drew closer to Earth. Strange phenomena began to occur during the week of March 20, when multiple observers reported seeing small sunward-pointing antitails on Bennett, along with jets and what several astronomers described as pinwheel streamers emanating from the nucleus. When the comet glowed at 1st magnitude on April 2, the American amateur astronomer John E. Bortle described "hoods" of light surrounding the comet's nucleus that were reminiscent of the strange, splayed coma of Comet Donati (C/1858 L1) – the comet viewed by Abraham Lincoln.

Comet Bennett showed unusual variation in its coma but also in the tails: The gas tail, in particular, changed dramatically over periods of a few days. For example, on April 4 observers noted it as patchy and irregular, while 3 days later it appeared straight and without irregularities. Bennett became distorted once again over the week that followed.

During the remainder of April Comet Bennett remained readily visible despite its slow fade. By month's end the comet's brightness dropped to magnitude 4.5, making it a somewhat faint naked-eye object, and the tail shrank to 20°. During May 1970 Bennett diminished to below naked-eye visibility and its tail shortened to about 9°, remaining visible in binoculars and small telescopes through August. Amateur astronomers continued observing Comet Bennett through year's end, by which time it faded to near-invisibility to hobbyists.

Elizabeth Roemer of the University of Arizona imaged the comet for the last time on February 27, 1971, and thereafter Comet Bennett was not seen again.

The next comet to make a splash in the popular culture was a would-be Great Comet, one that would fall short of astronomers' hopes and dreams. The Czech astronomer who found this comet, Luboš Kohoutek (1935–), is a prolific comet and asteroid discoverer and, among other things, coauthor of the well-known and cherished catalog of planetary nebulae known as the Perek-Kohoutek Catalog (actual title: *Catalogue of Galactic Planetary Nebulae*), published in 1967.

On March 18, 1973, Kohoutek found a fuzzy object on a photographic plate he had exposed at the Hamburg Observatory some 11 days earlier. Kohoutek described the object as condensed and having a diffuse envelope, and the comet moved slowly enough that it was easy to image again when he tried on March 21. Brian Marsden found that Kohoutek was near Jupiter distance from the Sun and that it promised to be exceptionally bright when close to the Sun at year's end and in January 1974. Astronomers were so emboldened by predictions they began to call Kohoutek the "comet of the century."

But as we've seen, Kohoutek would not cooperate fully with the hype. By the time the Czech astronomer Antonín Mrkos (1918–1996) observed the comet as it approached the Sun in late April, it glowed faintly at magnitude 14.5. By the first week of May the comet was lost in the Sun's glare, too close to be observed until it reemerged in late September.

The Japanese astronomer Tsutomu Seki spotted the comet on the other side of the Sun on September 23, glowing at magnitude 10.5 in the morning sky. Comet Kohoutek brightened over the coming weeks, and John Bortle estimated the comet at magnitude 8.3 by October 27. The comet brightened during November to an impressive object as viewed with telescopes and even binoculars, rising to magnitude 5.5 by month's end and sporting a tail some 4° long. Observers with dark skies could then just make out the comet as a naked-eye object.

Naked-eye comet though it was, Kohoutek failed to live up to anything close to "comet of the century." In December the comet rose to magnitude 4 but sank, day by day, steadily into the morning twilight. The amateur astronomer Walter Haas (1917–), founder of the Association of Lunar and Planetary Observers, traced out the comet's tail to a length of 18° on December 18. But the deepening twilight each morning obliterated the faint tail successively, even as the comet rose to 3d magnitude. Late in the month, only the astronauts aboard the *Skylab* space station could see the comet above Earth's limb. They could see Kohoutek as a magnificently bright object, viewing it with a coronagraph on the day of perihelion at magnitude –3.

By the time earthbound observers spotted Kohoutek after sunset in December, the comet glowed at magnitude –0.5 but was steeped in twilight. As the comet moved into a darker sky, it faded dramatically. By mid-January 1974 it had lost so much brightness it was a faint naked-eye object. By the end of January it was purely relegated to binocular and telescope users. As the comet swept past Earth, disappointingly faint, it nonetheless grew a 25°-long tail and throughout January displayed an antitail.

Kohoutek's performance simply could not match the hype. In February it faded to nearly 9th magnitude by month's end and displayed a diminutive tail of 1°. A few observations were made thereafter, but the anticipation was nullified and the party was over. The last observation of Kohoutek occurred on April 26, when Elizabeth Roemer imaged the comet and estimated its magnitude at 18. The orbit calculated by astronomers had misled them. Believed to be an Oort Cloud comet with abundant presumably fresh ice, Kohoutek instead proved, on the basis of telescopic and infrared studies, to be a Kuiper Belt object that may have been far rockier and much less icy than astronomers would have believed.

In the wake of Kohoutek, astronomers were very slow to become excited about the prospects for any comet, known or newly discovered. So the excitement built at a relative snail's pace when Danish astronomer Richard M. West (1941–) found a fuzzy object on November 5, 1975, on plates made with the 1-m Schmidt camera at the European Southern Observatory's facility in La Silla, Chile. West (pronounced "vest") found an object he described as between 14th and 15th magnitude and with a slight coma 2 to 3 arcminutes in diameter.

When Brian Marsden at the Central Bureau for Astronomical Telegrams received word of the discovery and calculated an orbit, he found the comet would become bright – naked-eye visible – by February 1976. So was announced Comet West (C/1975 V1). Marsden predicted a perihelion date of February 25, 1976, at which time the comet would be some 29.5 million km from the Sun. The comet's closest passage to Earth would occur on March 4, at a distance of 119.7 million km.

Comet West brightened rapidly as it approached Earth and the Sun. On November 25, 1975, Leo Boethin, a missionary living in the Philippines, became the first to spot West visually, glowing faintly at magnitude 12.7 and with a tail measuring just 4 arcminutes long. On December 1 the comet became visible to Northern Hemisphere observers, and Tsutomu Seki, at the Geisei Observatory in Japan, estimated the comet's magnitude at 12.5. Before the comet snuck too close to the Sun to remain visible, American astronomer Henry Giclas (1910–2007) at Lowell Observatory imaged West.

In the Southern Hemisphere, however, observations were still possible. Comet West brightened to 9th magnitude by the end of December. As the New Year rolled over into January, observations continued apace and by month's end the comet reached naked-eye visibility for Southern Hemisphere observers. Comet West's appearance became striking in February 1976, when the comet brightened to 4th magnitude (by midmonth) and then dramatically during the next 2 weeks, reaching magnitude –1.5 by February 24. Close to perihelion passage on February 25, a number of observers spotted Comet West in the daytime sky, using binoculars or telescopes.

After perihelion, the comet became visible to naked-eye viewers once again on March 1. Prominent in the very early morning sky, rising a short time before the

Sun, Comet West displayed a brilliant nucleus with a reasonably bright inner coma, both of which could be seen well in the twilit sky. The comet's tail stretched only about 2°, but the overall magnitude was impressively bright at about –1. Over the following days the tail grew substantially, however, to 10° on March 2 and as much as 30° long by March 8. West's tail also took on some interesting features; three components were easily visible in the fan-shaped dust tail, and the bluish gas tail remained sharply focused and narrow.

On March 4 Comet West made its closest approach to Earth, and the comet remained visible to naked-eye observers throughout March. On March 5 a large number of sky watchers armed with telescopes noted the comet's nucleus appeared to be split into two parts, separated by about 3 arcseconds. Amazingly, 6 days later telescopic viewers saw the nucleus as four distinct components (which were labeled with letters, as is the custom). Components A, B, and D remained gravitationally bound and were visible for months, whereas component C faded rapidly, disappearing by March 25.

As the nucleus of Comet West split apart, the comet faded. By March 31 the comet glowed at magnitude 4.7, keeping it a respectable naked-eye object, but the glorious brilliance of the comet was gone. By early April the comet stretched 4° long. Late April witnessed Comet West's sinking to the naked-eye limit with merely a 1° tail. Amateur and professional astronomers continued to observe Comet West with binoculars and telescopes for the 2 months following, and many unusual observations took place in May and June.

During early summer 1976 Comet West became fainter and more diffuse, making estimating the comet's magnitude more challenging. The comet faded to 8th or 9th magnitude by the end of June. Comet West became increasingly more difficult to observe in July and August, and the last verifiable observation was that of the amateur astronomer John Bortle on August 25, when the comet's magnitude was around 11.0. Photographers continued recording the comet until September 25, when Elizabeth Roemer of the University of Arizona made the last images of the comet, estimating the remaining portions of the nucleus at magnitudes 19.3 (A), 20.1 (B), and 20.3 (D).

We've already seen the extraordinary buildup – like nothing that preceded it – to the appearance in 1985 and 1986 of Halley's Comet (1P/Halley). Professional astronomers armed with ground-based telescopes and their armada of spacecraft set to encounter the comet produced a treasure trove of cometary science, as we've seen. But in no way did sheer excitement build for a group of hobby enthusiasts at any other time quite as it did during the run-up to Halley's Comet among the amateur astronomy community. In early 1985 commenced a year-and-a-half-long period of sheer bliss.

And the journey really began long before 1985. The initial spike of interest began in October 1982 when astronomers at Palomar Mountain Observatory first imaged the comet on its 1980s return to the inner solar system. Anticipation was so high that in the June 1983 issue of *Astronomy* magazine I wrote a story called "Waiting for Halley." Our readers were so enthused by the prospects of beginning to think and talk about Halley that I penned the line "It is now less than three years and counting until it will be visible in backyard scopes. And those years will pass more quickly than you might think."

Everyone knew that we had drawn a bum hand from the cosmos: Our apparition of Halley would be a particularly unfavorable one, because Earth would be on the other side of the Sun from the comet when Halley would be at perihelion. Despite the two relatively close approaches of Halley to Earth during this cosmic dance, in November 1985 and mid-April 1986, the comet would never get closer than 60 million km to Earth, some three times the distance of the comet's closest approach in 1910.

Scientifically exciting as it was, for amateur astronomers Halley would be psychologically electrifying but observationally a little bit of a letdown. Astronomers calculated a predicted brightness of about 4th magnitude at the comet's peak in April 1986. Adding to the disappointment for many observers was the fact that, at its best, Halley would be best visible from the Southern Hemisphere, while most observers live in the North.

Despite these limiting factors, the 1985–6 apparition of Halley's Comet was an immense party in which observers and astrophotographers celebrated their love of the skies through the comet. And because of that, I include a summary of the events here – despite the fact that, in a technical sense, Halley was not a Great Comet. In our lifetimes, it was only the most famous comet.

The anticipation of what backyard observers would see when Halley brightened built for months. The comet would presumably become visible for most observers who had reasonably large telescopes around mid-September 1985, when it would reach about 13th magnitude. At that time the comet would be near the star Eta (η) Orionis in our sky, and roughly equidistant from Earth and at the Sun, some 375 million km from each.

October 1985 would see the comet brighten considerably to within reach of most backyard telescopes. By mid-October Halley would brighten to about 10th magnitude and head toward the Pleiades star cluster (M45) in Taurus. Late in the month, at around 9th magnitude, the comet would slip past the Crab Nebula (M1), the two objects visible together in a low-power field of view.

The following month would see the comet further brighten considerably. Halley would then approach naked-eye visibility as it moved past the Pleiades. December would see the comet brighten to become a faint naked-eye object and remain visible low in the western evening sky. January 1986 would bring Halley at 5th magnitude and moving through the constellation Aquarius.

On February 9, 1986, Comet Halley would reach perihelion at a distance of some 87.7 million km from the Sun. Lost in the Sun's glare, the comet was expected to glow at around 4th magnitude during its peak brightness. At this time it would pass very close to the Saturn Nebula (NGC 7009) in Aquarius.

Observers would recover Halley, still predicted to glow pretty brightly, in the March morning sky. By March 7 the comet would move into Capricornus and glow at 5th magnitude. It would then slide past the globular star cluster M75 and make its way into Sagittarius, brightening slightly as it moved past Earth and resting at month's end near the Teapot asterism.

Halley would then plunge southward, disappearing for Northern Hemisphere viewers for a time until reappearing in the evening sky in mid-April. By that time the comet, glowing at around magnitude 4.5, would pass close to the peculiar galaxy Centaurus A (NGC 5128). Toward month's end the comet then would begin to climb northward as it slowly faded, starting its outward journey in the solar system. By May 1986 the comet would slip down beneath the naked-eye threshold as it moved through the southern constellations Crater and Hydra. Finally, lost in the sunset in June, the observational experience with the world's most famous comet would be over.

As I wrote back in 1983, though, astronomers were even then hedging their bets. "The comet may turn out to be brighter than we think," I penned, "or may be even dimmer than fourth magnitude. We can use the information on Halley from past observers, but we still must be prepared for the unexpected. All we can do for the moment is to plot out the estimated orbit on a star chart, polish up our eyepieces, and keep waiting."

As it turned out, the unprecedented observational experience with Comet Halley was, for amateur astronomers, a sensational event - despite the fact that the comet never brightened much beyond 3d magnitude, pretty much as predicted. After the comet's recovery in October 1982 at Palomar, the first CCD image of Halley's coma was made on September 25, 1984, when the comet was 912 million km from the Sun. On January 23, 1985, the amateur astronomer and *Astronomy* magazine columnist Stephen James O'Meara (1956–) made the first visual observation of the comet using the 24-inch telescope on Mauna Kea, Hawaii. The comet then appeared as a pinpoint nucleus glowing faintly at magnitude 19.6, surrounded by a tiny haze.

On November 8, 1985, another wave of excitement passed through the astronomical community when two experienced California observers, astronomers Stephen J. Edberg (1952–) and Charles S. Morris, glimpsed the comet with their eyes alone from a very dark site, seeing a faint, misty patch of light. Comet Halley remained visible to the naked eye until May 30, 1986, a remarkably long run of naked-eye visibility, except for the brief periods when it was very close to the Sun.

Halley's tail was first spotted on images made on July 12, 1985. In December 1985 and January 1986, the tail split into two components measuring 3° and 4° and was readily apparent. On January 9 astronomers witnessed a significant so-called disconnection event wherein a portion of the gas tail appeared to break free and separate from the part connecting it to the coma, causing a visual gap in the tail. If magnetic field lines are squeezed along the comet's tail, such disconnection events can occur in comets. Similar disconnection events with Halley took place on March 9 and April 11 and lasted for a couple of days each time.

After its perihelion passage, Halley was a bright naked-eye object in the predawn sky during late February and in March, although it was better seen in the Southern Hemisphere than in the North. The comet shone at about magnitude 2.6 at its very brightest and displayed a multifaceted tail stretching as long as 12°. Jets and fans emanated from the comet's nucleus, and the tail at times seemed fanned in a splay that consisted of multiple dust tails and even the appearance of a short, spiky antitail. Photographs taken on February 22 seemed to show these features at their height. Images taken at the Siding Spring Observatory in Australia demonstrated multiple fanned dust tails, as many as 13 distinctly seen at once. And on March 25 the comet showed a peculiar "kink," a bend in the tail that introduced a temporary, jagged right angle in the gas tail, separating from the dust tail some 3° from the nucleus.

Odd behavior aside, Halley's performance lived up pretty straightforwardly to the predictions. As it made its closest approach to Earth on April 11, the comet appeared to show a fatter, heftier coma, with a bright coma larger than the Full Moon and an impressively bright core. The tail was then geometrically altered to appear more or less fan-shaped, with a broad collar of light around the inner coma. Some observers had difficulty making out details in the tail at this time because the comet was floating in front of the brightest part of the summer Milky Way.

The morning of April 14, 1986, featured an unusual display from Halley, when it showed a tail that appeared to be separated into thirds. Soon thereafter the tail grew to some 10° in length, and then, for observers in the South, the tail stretched magnificently to nearly twice that length. Over the following days, Southern Hemisphere observers reported very long tails on the comet: The Australian amateur astronomer Terry Lovejoy (1966–) described a tail as long as 46° under a very dark sky and suspected an even longer, extremely faint extension.

During May Halley's Comet faded rapidly, right on cue, and the tail shrank to far less impressive stature. June 14 marked the last time the comet's gas tail was captured in a photograph. A persistent CCD imager recorded the last image of the comet's dust tail, amazingly enough, on April 1, 1987. The esteemed comet discoverer and astronomer David H. Levy was the last to see Comet Halley visually, on February 23, 1988, when he viewed it with the 61-inch telescope at Steward Observatory near Tucson, Arizona.

The last vestiges of the coma were captured by May 1988, and in February 1990 the Danish astronomer Richard M. West captured a CCD image of the comet revealing no more than a dim speck of light. Amazingly, Halley underwent an outburst and brightened in February and March 1991, when the comet was 2.1 billion km from the Sun. The comet brightened to magnitude 19, increasing its light intensity by a factor of 250. When astronomers imaged Halley with the Very Large Telescope in Chile on March 8, 2003, it glowed extremely dimly at magnitude 28, a barely visible pinpoint in the distant deep recesses of the solar system.

The early 1990s brought the appearance of a comet that was not spectacularly bright but was instead spectacularly interesting. Back in 1862 the American astronomers Lewis Swift (1820–1913) and Horace P. Tuttle (1837–1923) within 3 days of each other independently discovered a comet, which eventually was designated 109P/Swift-Tuttle.

After its discovery, astronomers found Swift-Tuttle has a period of about 133 years, and much later they calculated its size as 26 km in diameter. The comet's major distinction among amateur astronomers is that it's the parent body of the most popular annual meteor shower, the Perseids, which peak every August 11 or 12. The shower, which features some 100 meteors visible each hour, takes place as Earth intersects the comet's orbit, allowing meteoritic particles to fall into Earth's atmosphere, where they heat up, ionize, and cause streaks of light. The shower is named after the constellation Perseus because that star group hosts the radiant, the point from which the meteors appear to originate.

Although Comet Swift-Tuttle has a relatively long period, its orbit carries it reasonably close to Earth. Astronomers eagerly anticipated the recovery of Swift-Tuttle in 1992, and the Japanese amateur astronomer Tsuruhiko Kiuchi unwittingly spotted it on September 27, 1992. He reported the discovery of a comet to the National Astronomical Observatory in Tokyo, who were the first to suggest this could be Swift-Tuttle. At discovery, the comet glowed at magnitude 11.5. Brian Marsden at the Central Bureau for Astronomical Telegrams had calculated, but not published, an orbit that concluded a perihelion date of December 11, 1992, only a day earlier than the actual one, and 143.4 million km.

There was a surprising 17-day discrepancy in the recalculated perihelion date, now thought to be due to systematic errors in historical measurements. Astronomers then took notice that if the comet's next perihelion passage were off by that margin, then Swift-Tuttle might approach extremely close to Earth or even impact it. At a diameter of 26 km, this would be calamitous. (Recall the size of the K-Pg impactor in the Yucatán Peninsula, the dinosaur killer, was probably about 10 km.) A collision with Swift-Tuttle's nucleus could kill all human beings on the planet.

The controversy over the danger posed by Comet Swift-Tuttle caused the American amateur astronomer Gary W. Kronk to investigate past orbits of the comet. Kronk found that comets observed in 69 B.C. and A.D. 188 by Chinese sky watchers probably constituted past apparitions of Swift-Tuttle. On the basis of these suspicions, Marsden recalculated the comet's orbit and found it to be stable and not an impact threat to Earth over the next two millennia. The comet will pass extremely close to our planet, some 7.5 million km away, about September 15, 4479. However the future goes, Comet Swift-Tuttle is the largest solar system object that will repeatedly approach Earth.

When observers swung their telescopes toward Swift-Tuttle in October 1992, they found the comet's magnitude was actually closer to 9. By the beginning of November the comet rose to naked-eye visibility, displaying a tail between 1° and 2° in length. At midmonth the comet brightened to 5th magnitude and, a week later, displayed a tail nearly 7° long on CCD images made by backyard astronomers.

Astronomers from the Observatorie de Paris, using the 1-m telescope at the Pic du Midi Observatory in the French Pyrenees, observed a jet emanating from the comet's nucleus November 20–26 and from it calculated a rotation period for the comet of 2.9 days. Throughout December observers continued to watch Swift-Tuttle, and by March 1995 the last observation was made by astronomers at Siding Spring Observatory in Australia.

A cometary dry spell led to a big discovery in the summer of 1995. On July 23, while observing the globular star cluster M70, two observers in the southwestern United States each stumbled on a fuzzy blob near the cluster in their eyepieces. In Cloudcroft, New Mexico, astronomer Alan Hale (1958–) saw the comet, while 440 miles away, in the desert near Stanfield, Arizona, Thomas J. Bopp (1949–) also spied it. After confirming observations and the calculation of an orbit by Brian Marsden, Comet Hale-Bopp (C/1995 O1) was born.

Incredibly, at its discovery, the comet was a staggering 1.1 billion kilometers away – between the orbits of Jupiter and Saturn. Although at this distance most comets would appear as tiny specks, Hale-Bopp already showed a coma. Scottish-Australian astronomer Robert McNaught (1956–), at the Anglo-Australian Telescope at Siding Spring, found a 1993 image of the comet showing a coma at the very great distance of 2 billion km, between the orbits of Saturn and Uranus. At that distance, Halley's Comet would have been more than 100 times fainter than Hale-Bopp.

Astronomers, professional and amateur, would be able to view Hale-Bopp for a long time before it moved into the inner solar system. They were immediately excited by the fact that Comet Hale-Bopp was likely to brighten substantially, reaching a perihelion distance of 136.7 million km on April 1, 1997, and swinging past Earth at a distance of 196.7 million km on March 22, 1997.

No one knew it quite yet at discovery, but Hale-Bopp was destined to become the brightest comet of its era and was perhaps the most widely observed comet in history, given the huge numbers of the general public who looked at it with unaided eyes, binoculars, or telescopes. Calculations showed the comet would peak shining brighter than any star in the sky except for Sirius, the sky's brightest star. Starting May 20, 1996, Hale-Bopp would remain visible to the naked eye for 569 days – more than 18½ months, the longest period of unaided visibility of a comet known. That record would double the previous 9-month record set by the Great Comet of 1811.

They also marveled over the comet's current 2,520-year period and wondered over the possible connection of an Egyptian comet observation to Hale-Bopp. The Egyptian pharaoh Pepi I reigned from 2332 to 2283 B.C., and Pepi I's pyramid at Saqqara, Egypt, contains a hieroglyph referring to a "nhh-star" (*nhh* is the word for "long-haired"). Comet Hale-Bopp could have appeared in Earth's skies during 2213 B.C. It could be that the Egyptians beat Hale and Bopp to the discovery by more than 4,200 years.

Astronomers relished the run-up to Hale-Bopp's period of brightness as they had never before with a Great Comet. In August 1995 the comet shone at magnitude 10.5, with a coma spanning 3 arcminutes and a hint of a tail. At this distance, the comet's diameter and prototail suggested an enormous physical size for the comet. (Astronomers would later determine the comet's nucleus stretched about 60 km across, several times the size of Halley's Comet.) By November 1995 Hale-Bopp had brightened to 10th magnitude but was slipping into the glare of twilight. During the first days of the New Year the comet slid just 2° from the Sun.

In February 1996, Comet Hale-Bopp reemerged. The Australian comet hunter Terry Lovejoy described the comet as "well condensed with a noticeably fan shaped coma and is significantly brighter than [it was] last year." During this period another bright comet, Hyakutake, received the lion's share of the attention, and few observers concentrated on Hale-Bopp. By late April 1996, however, Hale-Bopp brightened nicely to within easy binocular range at 8th magnitude.

On May 20, Terry Lovejoy became the first observer to spy Comet Hale-Bopp with the eye alone. He then used a pair of 10x50 binoculars to estimate the comet's brightness at magnitude 6.7 and reported a coma stretching 15 arcminutes in diameter – half the width of the Full Moon. By early July, Hale-Bopp reached magnitude 5.5 and was widely visible as a naked-eye object.

Then came the unexpected. During July Hale-Bopp seemed to be frozen in place at about the same brightness. Both professional and amateur astronomers began to think the comet could be repeating the embarrassing legacy of Comet Kohoutek, the "comet of the century" that fizzled. August and the first couple of weeks of September did little to avert the uneasy feeling, which in some amateur circles amounted to a small panic. By September 21, when the comet was within 500 million km of the Sun, it began to brighten again. The comet inched upward to about magnitude 5.3 by month's end.

Despite the comet's sluggish performance, the professional community basked in a bundle of data collected on the comet. It was the best view astronomers had ever had of a comet that was still so distant. In hindsight, astronomers concluded that the strange behavior of a halt in brightening was probably normal at such a large distance from the Sun, and that they had simply not seen it in so much detail with previous comets.

During the final weeks of 1996 Comet Hale-Bopp established itself as one of history's Great Comets. It rose to 5th magnitude by Halloween and 4th magnitude by mid-December. By this time the comet had approached to within 300 million km of the Sun, twice the orbital distance of Earth, and slinked into the twilight as it lay near the other side of the Sun from our planet.

January 1997 witnessed Hale-Bopp still situated in morning twilight for those in the best position, the Northern Hemisphere. On January 6 observers reported magnitude estimates centered around 3.2, most of them from those situated at high northern latitudes. The most southerly observer to record observations of the comet at this time was, ironically enough, Alan Hale himself, who spotted the comet from New Mexico on January 2 and estimated its magnitude at between 3 and 3.5. The morning sky favored a larger number of observers after about January 10, as the comet's apparent distance from the Sun increased.

Now the stage was set for Comet Hale-Bopp to put on its spectacular show. The comet swung into the evening sky and brightened rapidly, reaching 2d magnitude in early February and rising to an impressive magnitude 0 by March 7, with a tail stretching 10°. By mid-March Hale-Bopp brightened further to magnitude –0.5, surpassing the predicted magnitude, and by March 20th reached magnitude –0.8. The incredible display of this amazingly bright comet and its prominently fanned gas tail awed millions of people across the globe, and the media coverage of the comet was enormous. A bizarre twist occurred on March 26 when 39 members of a religious cult in California called Heaven's Gate committed suicide, believing that somehow the comet's appearance would convey them into a magical afterlife.

Comet Hale-Bopp's perihelion date arrived on April Fool's Day, and reliable magnitude estimates then ranged as high as –1.3 with a tail stretching as much as 20°. The amazing gas tail (strongly blue in photographs) and wide, fanned dust tail each measured most of this length, and some observers saw one or the other as slightly longer on different nights. But as April matured the gas tail faded in prominence, spreading out in width.

May 1997 saw Hale-Bopp dropping in declination so that it moved into the Southern Hemisphere sky. Although the comet began the month shining at magnitude 0.3, it faded to 2d magnitude by month's end. In June and July twilight again interfered with observing the comet. During the late summer the comet dropped in brightness to 4th magnitude, and by October it was no longer a naked-eye object for many observers. Noting an outburst in December, however, Terry Lovejoy saw the comet at magnitude 6.8 in a dark sky on December 10.

Comet Hale-Bopp faded to 13th magnitude by 1999 and 16th magnitude by 2003, leaving in its wake one of the most memorable experiences with a comet in recorded history.

Just as Comet Hale-Bopp was in midstream, heating up as a binocular object on January 31, 1996, the Japanese amateur astronomer Yuji Hyakutake (1950–2002) was sweeping the sky with a pair of 25x150 binoculars. Hyakutake spotted a fuzzy object that appeared diffuse and had a slight central condensation, the whole object glowing at around magnitude 11. The comet's coma spanned about 2.5 arcminutes. Astronomers followed up with observations, and soon the object was designated Comet Hyakutake (C/1996 B2). Incredibly, this was the second comet found by Hyakutake in 5 weeks and lay just 4° from the location where his first comet, C/1995 Y1, was spotted.

Astronomers calculated Hyakutake's date of perihelion as May 1, 1996, when it would pass just 34.4 million km from the Sun. More impressively, the comet would pass Earth at a close distance: 15 million km, on March 25, 1996. Thus, it would become the thirty-third closest comet known to have passed by Earth. The comet's diameter was subsequently found to be 4.2 km, its rotation period some 6 hours, and the orbital period 17,000 years – but the 1996 encounter perturbed the orbit such that it will now return again in some 70,000 years.

On February 1, 1996, observers around the globe commenced observing the comet, and magnitude estimates ranged from 9 to nearly 12. By midmonth the comet brightened to the point of binocular visibility, at a magnitude around 8.5 with a coma measuring 8 arcminutes across. The comet reached naked-eye visibility around the end of the month, and Australian comet hunter Terry Lovejoy made the first verified naked-eye sighting of Hyakutake. The comet then glowed around 6th magnitude and sported a tail some 1° long.

Comet Hyakutake became a spectacular comet during March 1996. On March 6 Terry Lovejoy and American observer Charles Morris estimated the brightness at about 5th magnitude, while on March 10 Gary Kronk spotted the comet close to the Moon while only using a pair of binoculars. By the middle of March the comet had become a respectable naked-eye object, shining at about magnitude 4.

The comet brightened dramatically during the middle of March, with observers reporting a range of magnitudes all centered on 2.8 on March 18, and most sky watchers described seeing a tail stretching 2° to 4° long. Hyakutake's coma now spanned the diameter of the Full Moon and the comet became a media sensation, as the public at large ventured outside to gaze at the celestial visitor. By March 22, Gary Kronk reported the comet at magnitude 1.4 and with a tail some 15° long.

As the comet moved toward and passed the date of closest approach, March 25, it became a nearly unprecedented spectacle. Magnitude estimates at the time pegged Hyakutake at –1 to 1st magnitude, depending on the observer, although it was probably on the brighter end of that range. The most incredible aspect of the comet's appearance was that the close passage to Earth and the geometry involved gave us an enormously long tail – which ranged from 45° to 70° depending on the observer's sky darkness. On March 26, the experienced observers Stephen James O'Meara and James V. Scotti (1960–) estimated the tail length at 100°, and observers everywhere marveled at the incredibly long, glowing, gossamerlike streamers of the comet's tail.

In late March Comet Hyakutake began to display some unusual phenomena. Astronomers at Pic du Midi and at Lowell Observatory in Arizona captured data that allowed estimation of the rotation period of Hyakutake's nucleus, at close to 6 hours. Planetary scientists used the 230-m Goldstone radio antenna to measure the nuclear size in the range of 4.2 km. Observations beginning on March 23 showed that small blobs of material had detached from the comet's nucleus; these were imaged by the American astronomer Harold Weaver and others using the Hubble Space Telescope. Curiously, astronomers on the imaging team concluded the blobs were made up of large dust particles because they did not believe fragments of the nucleus would be accelerated in the tail, whereas these blobs were.

In April, Comet Hyakutake faded but observers still had a field day with it. Early in the month the comet glowed around 2d magnitude and displayed a tail still stretching at least 15° long. A total eclipse of the Moon took place on April 4 and sky watchers spotted the comet during the eclipse as the sky darkened (the bright Moon otherwise had interfered for a number of days). By the end of April's first week the Moon moved away and magnitude estimates again placed the comet at around 2d magnitude. Observers with very dark skies still estimated tail lengths of as much as 50°; Stephen James O'Meara proposed tail lengths of 60° on April 8 and 90° on April 9, and the comet passed close to Venus on April 11.

The comet put on a peculiar show on April 14, when observers noticed changes in its coma. Suddenly the coma seemed to have a brighter and more yellowish center. The comet brightened overall by a third of a magnitude. In Illinois, Gary Kronk observed the comet and spied a jet protruding from the nucleus on April 16, but a day later it was gone. Instead, he saw a "cloud" extending northward from the nuclear area. And a "spike" of glowing material emanated into the cloud from the position of the nucleus.

Comet Hyakutake now moved increasingly into twilight and was harder to spot. By April 21 magnitude estimates were so scattered that they varied from 1.6 to 4.1. After April 25, the comet became very difficult to see. With perihelion approaching, John Bortle spotted the comet for the last time on April 28, just 4° off the horizon, and glowing at about 3d magnitude with no tail. The comet was a mere 12° away from the Sun at this time and Bortle found it with a pair of 15x80 binoculars.

On May 1, 1996, the very day of perihelion, the newly commissioned *Solar and Heliospheric Observatory* (*SOHO*) spacecraft imaged Comet Hyakutake. For most observers on Earth, who reside in the Northern Hemisphere, the postperihelion showing of Hyakutake was not favorable. The comet slinked above the morning sky twilight only as the sky brightened appreciably each day. But Southern Hemisphere viewers had a better angle: Australian comet hunter Gordon Garradd was the first to see it, on May 10. Visible in bright twilight, the comet appeared to glow at about 3d magnitude and showed no discernible tail. By the following day several observers picked it up as a naked-eye object once again.

On May 13 sky watchers across the Southern Hemisphere widely reported seeing the comet at around 3d magnitude. A week later the comet's brightness had faded to magnitude 4.5, and the tail stretched a mere 7° long. Again, the comet was now placed in a dark sky, away from twilight, but now its days were numbered. By late June the comet dropped to below naked-eye visibility, and it continued fading through the summer. Gordon Garradd made a final observation of the comet on October 24, 1996, when the fuzzy object glowed meekly at close to magnitude 17. The brilliant and historic apparition of Comet Hyakutake was over.

On February 1, 2002, three observers in widely different locations found a fuzzy object in the evening sky. Kaoru Ikeya of Comet Ikeya-Seki fame, some 37 years after his most famous find, spotted a cometary object using a 10-inch reflector from Mori, Shizuoka, Japan. Some 30 minutes later comet hunter Daqing Zhang (1969–), using an 8-inch telescope, found the same object while observing near Kaifeng, Henan Province, China. A third independent discovery, by Brazilian amateur astronomer Paulo Raymundo, was made too late for recognition – the comet was confirmed and designated 153P/Ikeya-Zhang.

Astronomers realized the orbit made it a periodic comet with a long period of 367.2 years. They also traced the comet's history backward and realized that the bright comet observed and remarked on by the Polish astronomer Johannes Hevelius (1611–1687) in 1661 (C/1661 C1) was the same object. They also saw that the comet would probably become the brightest since Hale-Bopp, and amateur astronomers readied for action.

After its discovery, the comet brightened nicely, but it remained at a low altitude and steeped in twilight enough to go under the radar for many observers. Late in February the comet reached naked-eye brightness, and at month's end observers were reporting magnitudes as high as 5, and some observers believed the comet experienced a minor outburst.

March brought a more pronounced appearance of Comet Ikeya-Zhang, which was now impressive in binoculars and displayed a tail stretching about 2°. In early

Figure 4.1. Believed to be a minor planet at its discovery, Comet LINEAR (C/2006 VZ13) developed a coma and showed itself as a comet. In this image made July 24, 2007, the comet gracefully passes globular cluster M3 in Canes Venatici. The imagers used an 8-inch f/2.75 astrograph, a CCD camera, and stacked exposures. Credit: Michael Jäger and Gerald Rhemann.

March the comet displayed a series of tail disconnection events. Late in the month most observers estimated the comet's brightness at about 3.5, and the tail length as as much as 4°. Ikeya-Zhang held up well throughout April 2002 and faded to below naked-eye visibility during May.

In 2003 and 2004 two comets surprised amateur astronomers with their impressive, unexpected performances. The LINEAR team announced the discovery of an asteroidlike object glowing dimly at magnitude 17.5 on images made October 14, 2002. Subsequent observations showed the object had a tiny, diffuse halo, and when Brian Marsden calculated the orbit, he found it to have a perihelion date of April 23, 2004. The object became Comet LINEAR (C/2002 T7).

During 2003 Comet LINEAR brightened slowly, transforming from an object suited for large amateur telescopes to a binocular object by year's end. In early 2004 LINEAR gave observers a Hale-Bopp-like scare when it brightened very slowly relative to the predictions. By March, however, twilight observations allowed astronomers to estimate the comet's magnitude at 7, and after conjunction with the Sun, the comet reemerged and was considerably brighter, achieving magnitude 4.6 by April 9.

The comet's perihelion came and went on April 23, but a close passage of Earth at a distance of only 40.4 million km on May 19 kept the comet's brightness slowly growing throughout May. Late in the month the comet peaked in brightness at magnitude 2.5 and was best placed for Southern Hemisphere viewers. Strangely, Comet LINEAR appeared to fluctuate in brightness throughout the month, as a result of nuclear activity. Most observers reported seeing a tail as long as 2°, while others reported in dark skies a tail three times as long. Late in the month the tail apparently

Figure 4.2. Comet NEAT (C/2001 Q4) shows a nearly forked tail in this image taken June 6, 2004, with a 102 mm refractor at f/6, a CCD camera, and stacked exposures. Credit: Paulo Candy.

increased enormously in size, perhaps to as much as 43° long, and then shrank suddenly. This comet began as a tiny speck of light that was suspected to be an asteroid and ended as a misbehaving mystery.

Just a month after the discovery of LINEAR, on November 7, 2002, astronomers using the 1.2-m Schmidt telescope at Haleakala Observatory on Maui, Hawaii, discovered a comet as part of the Near Earth Asteroid Tracking (NEAT) program. Designated Comet NEAT (C/2002 V1), the object would pass closest to Earth on December 24, 2002, at some 119.7 million km, and reach perihelion on February 18, 2003, at a distance of 15 million km. Because it would pass so close to the Sun at perihelion, astronomers expected this comet would likely break apart.

The comet commenced the month of December 2002 glowing at 12th magnitude, rising to 8th magnitude by month's end. Astronomers believed the comet's rapid rise in apparent brightness could be attributed to a small nuclear size. Noting the comet's intrinsic faintness, John Bortle advised that the nucleus might not survive the close encounter with the Sun. In January 2003 observers saw it brighten to near-naked-eye visibility, but bright moonlight made estimates difficult.

In February NEAT's brightening slowed; most observers estimated the magnitude at 5, with a 2° tail. But the comet's coming close encounter with the Sun would cause it to brighten dramatically. In strong twilight, observers estimated the brightness at magnitude 3.5 on February 10 and 2.0 on February 13, with an outburst apparently occurring on February 7 and 8. For several days beginning on February 16, the *SOHO* spacecraft imaged Comet NEAT. Amazingly, the small comet held together and slinked back toward the outer solar system, having provided another fine, brief comet for amateur astronomers.

On March 23, 2004, one of the grand old masters of comet hunting, the Australian amateur William A. Bradfield, found his 18th comet. Bradfield had been sweeping

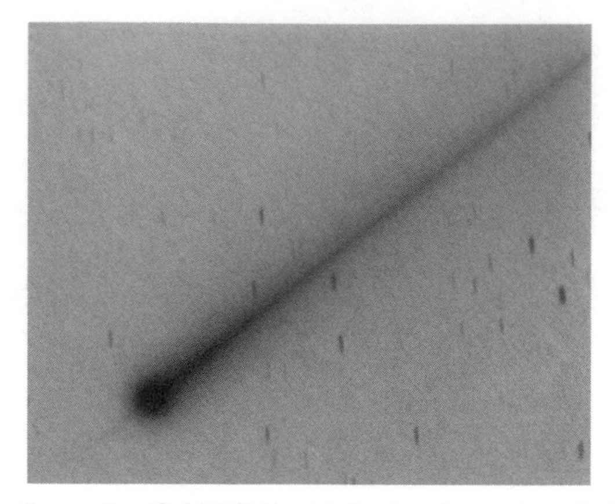

Figure 4.3. Comet Bradfield (C/2004 F4) displayed a weak antitail during its final days of brightness, as captured here on May 3, 2004. The imager used a 160 mm astrograph, a CCD camera, and stacked exposures. Credit: Dennis Persyk.

the sky along the twilit horizon, looking for sungrazers. Bradfield recovered the comet 2 weeks later, and subsequently Rob McNaught also observed it, leading to magnitude estimates of about 5. On April 12 Terry Lovejoy estimated the magnitude as 3.3. The object became known as Comet Bradfield (C/2004 F4) and the early orbital calculations suggested a perihelion date of April 17 at a distance of 25.3 million km (Figure 4.3).

Just 5 days past perihelion, Austrian amateur astronomers Michael Jäger and Gerald Rhemann found the comet bathed in deep twilight. The *SOHO* spacecraft had observed it through the days before and after perihelion, and the comet then shone at around 2d magnitude. A short time after the Austrian observations, Alan Hale in New Mexico observed Comet Bradfield and estimated its brightness at magnitude 4.5. Many observers saw Comet Bradfield in the final days of April in the morning sky, and the comet sported a surprisingly long tail of some 10° before it faded.

The next big comet surprise occurred on August 8, 2006, when Australian Rob McNaught of the Siding Spring Observatory imaged a fuzzy blob on plates made with the 0.5-m Uppsala Schmidt telescope. The comet then glowed dimly at 17th magnitude and displayed a tiny coma stretching 20 arcseconds. When Brian Marsden calculated the orbit, he found the object would reach perihelion on January 12, 2007, at a distance of 25.6 million km from the Sun. The world now welcomed Comet McNaught (C/2006 P1) into the fold. The comet was destined to be the second brightest since 1935 but was predominantly visible in the Southern Hemisphere, such that most northerners had only a brief glimpse at it.

During September professional astronomers took note of Comet McNaught, following it closely, as amateurs armed with very large telescopes joined the queue. By September 22, observers reported the comet glowing at magnitude 13.3 with a

Figure 4.4. Comet McNaught (C/2009 R1) displays a dazzling coma and a tail containing multiple streamers in this image made June 10, 2010, with an 8-inch f/3.6 astrograph, a CCD camera, and stacked exposures. Credit: Gerald Rhemann.

coma spanning 1.5 arcminutes. As October unfolded and McNaught headed toward the Sun, it became more difficult to observe. In November most observations were steeped in twilight. Near midmonth the comet approached 9th magnitude in brightness. December saw the comet hopelessly lost in twilight; however, observers spotted it again on December 26, estimating its magnitude at 4.5. By now the comet had developed a diminutive tail of 4 arcminutes length.

As the calendar rolled into 2007, the comet brightened rapidly, still immersed in twilight. Magnitude estimates ranged from 3 to 1.5. By the second week of January, millions of people began to see the comet low in the evening sky, causing the biggest dose of comet fever since the days of Hale-Bopp. By January 6 the comet shone at magnitude 0, but just 5 days later the comet had dramatically brightened to magnitude –3. During these few days, numerous Northern Hemisphere observers marveled at the comet as it sank deeper into the western twilight. It made a very impressive sight despite being bathed in bright twilight.

As the comet raced toward perihelion, it sank out of sight for most observers but was picked up by satellites including *SOHO* and the Solar Terrestrial Relations Observatories (STEREO). As these observatories were taking data, on January 13 and 14, Comet McNaught reached its peak brightness, at magnitude –6. Observers spotted the comet in broad daylight by blocking out the nearby disk of the Sun.

McNaught then moved into the evening sky as viewed by Southern Hemisphere observers, who were treated to a spectacularly detailed, curving, multifaceted dust tail. The comet's appearance brought back memories of Comet West in 1976. As the comet faded to magnitude 0 and climbed out of the twilight, its tail stretched as long as 24° in late January and showed multiple luminous bands. Observers continued to

Figure 4.5. Comet Lovejoy (C/2007 E2) appeared as a condensed coma and faint, wispy tail when it was imaged on May 11, 2007, using a 6-inch refractor, a CCD camera, and stacked exposures. Credit: John Chumack.

see the comet in February and March, and it faded to about the naked-eye limit by about the end of the first week of March.

Although it was chiefly a treat for the Southern Hemisphere, Comet McNaught was clearly the most impressive sky treat since Hale-Bopp.

Late in 2011, the prolific Australian comet watcher Terry Lovejoy bagged a sensational object that he found as a fuzzy blob on November 27 of that year. When he found it with an 8-inch Schmidt-Cassegrain telescope equipped with a CCD camera, the comet glowed at 13th magnitude. Confirmation and an orbit determination followed quickly and the object became Comet Lovejoy (C/2011 W3). Astronomers calculated a perihelion date of December 16, 2011, at a distance of just 830,000 km from the Sun. Comet Lovejoy was a Kreutz Sungrazer, the first to become visible from Earth since 1970. Astronomers found the comet has a period of about 681 years.

Southern Hemisphere viewers focused in on the comet beginning on December 2. The following day, Lovejoy estimated the comet's magnitude at 11.6 and noted the coma spanned 1 arcminute. Astronomers questioned whether or not Lovejoy would break up through the force of the Sun's gravity on its approach to perihelion. The comet's initial brightness suggested it might be a physically small object. But the comet passed perihelion, observed by *SOHO* and other instruments, and actually seemed to brighten in the days immediately afterward.

After perihelion, California sky watchers picked up the comet again on December 17. Soon thereafter, Lovejoy estimated the comet's brightness at magnitude –1.2. As the comet moved away from the Sun, numerous observers watched it from all over the world. On December 17 observers estimated the comet's brightness at –2.9. Three days later, the comet transformed itself. The nucleus now suddenly seemed to be shaped like a bar. On this day several observers estimated the comet at magnitude

Figure 4.6. Comet PANSTARRS (C/2011 L4) rises above the city of Bariloche and Mt. Catedral, in Argentina, on March 3, 2013. The imager used a 100 mm lens at f/2 at ISO 800 and a 2.5-second exposure. Credit: Guillermo Abramson.

2.0, and the tail now stretched 15° long. Other observers saw the tail to be some 38° long as the comet faded to 5th magnitude by year's end, and soon the comet was just a memory.

And that takes us to the year 2013. The first of the two comets that were becoming bright as I was writing this text was discovered in 2011. The 1.8-m Pan-STARRS Telescope at Haleakala, Maui, Hawaii, found this object on June 6 of that year, when the object's magnitude was about 19.5. (Pan-STARRS is an acronym for Panoramic Survey Telescope and Rapid Response System.) When discovered, the object was about 1.2 billion km from the Sun. Astronomers confirmed the object as a comet and designated it Comet PANSTARRS (C/2011 L4) (Figure 4.6). The perihelion date would be March 11, 2013, at a distance of 44.9 million km from the Sun.

The comet brightened to binocular range by late 2012 and reached naked-eye visibility just as I was writing this, on the first days of March. PANSTARRS passed closest to Earth on March 5, at a distance of 164.6 million km, and continued to brighten beautifully as it passed perihelion. As I wrote this, observers were enjoying PANSTARRS and keeping the faith in the comet they hope will be much brighter yet – Comet ISON (C/2012 S1).

5

Comets in Human Culture

Comets have always been afforded a special place in the minds of human beings. In fact, from the earliest days of recorded history up through the 17th century, most people thought they were harbingers of doom or portents or some incredibly important events - perhaps good but usually bad. Thought to be specters from beyond, ghosts sent from heaven or hell and carrying wickedness or good, comets spent the entire history of human culture being misunderstood until just the last 300 years. And the real understanding of comets, as we've seen, has only developed over the past century. It's difficult to blame the ancients for singling out comets, as so few things in the sky seem to change over short time intervals. To most, the sky, filled with stars, seems essentially static. Even the motions of the planets, the Moon, and the Sun are comfortable. But comets were outside the realm of the comfortable, and their uninvited, sudden appearances would shake the foundations of belief.

The Roman philosopher Lucius Annaeus Seneca, aka Seneca the Younger (*ca.* 4 B.C.–A.D. 65), summarized the early take on comets - as opposed to the always-visible parts of the heavens, the Sun, Moon, and so on - in one of his writings. "No man is so utterly dull and obtuse," he penned,

> with head so bent on Earth, as never to lift himself up and rise with all his soul to the contemplation of the starry heavens, especially when some fresh wonder shows a beacon-light in the sky. As long as the ordinary course of heaven runs on, custom robs it of its real size. Such is our constitution that objects of daily occurrence pass us unnoticed even when most worthy of our admiration. On the other hand, the sight even of trifling things is attractive if their appearance is unusual. So this concourse of stars, which paints with beauty the spacious firmament on

> high, gathers no concourse of the nation. But when there is any change in the wonted order, then all eyes are turned to the sky. … So natural is it to admire what is strange rather than what is great.

"The same thing holds in regards to comets," continued Seneca. "If one of these infrequent fires of unusual shape have made its appearance, everybody is eager to know what it is. Blind to all the other celestial bodies, each asks about the newcomer; one is not quite sure whether to admire or fear it. Persons there are who seek to inspire terror by forecasting its grave import. And so people keep asking and wishing to know whether it is a portent or a star."

The first organized thoughts about comets, in an analytical sense, grew out of the work and associations of the Greek philosopher Aristotle (384 B.C.–322 B.C.), a student of Plato's and teacher of Alexander the Great. In his *Meteorology* (*Meteorologica* in Latin), Aristotle laid down his views about the geological sciences. Not only did he define the four basic elements – fire, water, air, and earth – he also discussed weather phenomena, earthquakes, the physics of moving bodies, hydrology, and comets. His text is also valuable as a review of previous ideas about comets, which he summarily dismissed.

Aristotle reviewed the notions of Greek philosopher Pythagoras of Samos (*ca.* 570 B.C.–*ca.* 495 B.C.), creator of the Pythagorean theorem, who fancied the motions of celestial bodies following elegant curves, held in highest reverence in the form of a circle. This notion subsisted until Kepler's laws of planetary motion came along in the 17th century. Aristotle summarized how Pythagoras and his followers believed that only one Great Comet existed and that all sightings of comets represented seeing the same object in different places at different times as it rose a short distance above the horizon.

The story was very different according to Greek mathematician Hippocrates of Chios (*ca.* 470 B.C.–410 B.C.), who supposed that a comet formed its tail by drawing water vapor away from Earth. The visibility of a comet's tail, he further believed, was dependent on the humidity of the region from which the comet was being viewed.

And even more strangely, Greek philosopher Anaxagoras of Clazomenae (*ca.* 500 B.C.–*ca.* 428 B.C.), along with his compatriot philosopher Democritus of Abdera (*ca.* 460 B.C.–*ca.* 370 B.C.), surmised that comets were formed by conjunctions of planets approaching each other in the sky and therefore appearing to contact, causing unusual phenomena like tails.

Dismantling these previous notions, Aristotle then laid forth his own ideas in the important work *Meteorology*. And they were incredibly influential, as Aristotle's concepts and ideals formed the architectural basis of many basic scientific thoughts for centuries. According to Aristotle, the notion of comets grew out of his celestial spheres, containing the basic forms of matter, which were concentric and ordered in density. The densest, Earth, lay in the center. Then came water, which was less dense but also heavy and relatively cold, like Earth. Next was the airy sphere, and above it the fiery sphere – which didn't consist of flames, but objects laden with the potential of fire. This distant storehouse contained flammable materials that, if agitated, would ignite and cause luminous phenomena. He later added a fifth element, ether or "quintessence," to be the outermost sphere, a changeless region that moved about in circles.

In this celestial universe of Aristotle's, comets formed when the Sun or even planets warmed Earth, thus causing warm, dry emissions from places on Earth to evaporate and rise upward. Simultaneously, cool, moist air evaporated and similarly rose upward. The cool, moist air interacted with the warmer air rising through it, and at the border of the fiery sphere, friction between the two columns ignited them, producing a comet. The comet's motions and behavior in the sky depended on the nature and amounts of the emissions from below.

Aristotle believed that comets foreshadowed atmospheric events on Earth like winds and droughts, reflecting those conditions under which they formed. Despite this adherence to promoting comets as local atmospheric phenomena, Aristotle proposed rational ideas based on large amounts of empirical observation. He turned out to be simply wrong. But the basic ideals of what Aristotle believed about comets lived on for many hundreds of years – longer than the cosmos of concentric spheres and circular orbits he believed in, which was soon smashed by the more sophisticated ideas of epicyclic orbits by Greco-Roman mathematician Claudius Ptolemy (*ca.* 90–*ca.* 168).

In the Roman world, more than 300 years after Aristotle's death, Seneca wrote about comets and much more in his famous work *Natural Questions* (*Naturales quaestiones*, in Latin), written about the year 63. And he wrote about them under considerable pressure from Emperor Nero (37–68), who had begun his life of wretched excess, total immorality, and criminality – and who had already arranged for the deaths of his mother, his wife, and his stepbrother. An imperial adviser, Seneca was accused of embezzlement and wrote in a style baldly flattering the emperor.

Seneca mentioned recent comets, recalling "one which appeared during the reign of Nero Caesar – which has redeemed comets from their bad character." He later referred to "the recent one which we saw during this joyous reign of Nero." The flattery worked only for a short time. Accused of complicity in the Pisonian conspiracy, a plot to kill Nero, Seneca was forced to commit suicide. Nonetheless, his writings about comets published in *Natural Questions* continued to influence thinking about the celestial visitors for many years to come.

Seneca believed that comets were not fiery apparitions but permanent creations of the natural world that would last and be seen for short periods because of their movements. He mentioned the observation of a comet during a solar eclipse and

Figure 5.1. In 2012 Comet Garradd (C/2009 P1) showed a beautiful antitail as it passed the background galaxy NGC 6339 in Hercules. This image was shot January 31, 2012, using a 12-inch f/3.6 astrograph, a CCD camera, and stacked exposures. Credit: Gerald Rhemann.

therefore believed many comets may move close to the Sun in the sky and be hidden by its glare. Impressively, Seneca proposed that comets move in circular orbits and travel around the sky, becoming invisible when they move behind planets. One of the chief arguments for comets' being atmospheric, and not distant celestial bodies, was the fact that their appearances and motions differed greatly from those of stars and planets.

For this, however, Seneca had a rational retort. "Nature does not turn out her work according to a single pattern," he wrote. "She prides herself upon her power of variation.... She does not often display comets; she has assigned them a different place, different periods from the other stars, and motions unlike theirs. She wished to enhance the greatness of her work by these strange visitants whose form is too beautiful to be thought accidental."

Seneca then looked forward to a distant future, realizing that understanding would deepen over the ages. "The day will yet come when posterity will be amazed that we remained ignorant of things that will to them seem so plain. ... Men will some day be able to demonstrate in what regions comets have their paths, why their course is so far moved from the other stars, what is their size and constitution. Let us be satisfied with what we have discovered, and leave a little truth for our descendants to find out."

As Roman philosophers went, Seneca was an impressive scientist in the making. He stood up for careful, analytical observations, and even though his conclusions were primitive, the analytical component was a big step forward. In fact, his belief in the periodic orbiting of comets caught on, finally, with Edmond Halley – more than 16 centuries after Seneca's death.

During Seneca's lifetime, however, another Roman influenced contemporary thinking on comets. The philosopher, author, and naturalist Gaius Plinius Secundis (23–79), better known as Pliny the Elder, was a close friend of the emperor Vespasian. A prolific writer, Pliny the Elder wrote numerous books and had many adventures before dying in a rescue operation during the famous eruption of Mount Vesuvius that overwhelmed Pompeii and Herculaneum.

Pliny the Elder's last work is *Natural History* (*Naturalis Historia* in Latin), written in 37 books and completed near the end of his life. It's also Pliny's only work to have survived. The book treats botany, zoology, metallurgy, mineralogy, and other assorted topics – including comets. Pliny's scope was far-reaching, and he told others he wanted to compile a book that would provide a general description of "everything that is known to exist through the Earth." Hooray for modest goals.

Because this work was so voluminous, its influence spread across the globe and lasted deep into the Middle Ages. But Pliny gathered all manner of information for his work, perhaps driven by its sheer heft, including nonsensical superstition, pseudoscience, incantations, and magic spells in the book as well as "balanced" ideas of the day. Leaning heavily on Aristotle, Pliny completely ignored Seneca and his "modern" views of comets. Instead, he delivered a barrage of stream-of-consciousness facts, sometimes in contradictory fashion.

Pliny wrote that some comets move and others don't. He suggested they could show up anywhere in the sky, but the ones in the south lacked a tail. He recorded the longest and shortest apparitions of a comet in the sky were 80 days and 7 days. He suggested that comets are most often observed in the northern sky and are often associated with the glow of the Milky Way. Pulling backward against reason, he fully supported the idea of comets as portents of doom and even provided something of a user's guide in determining what types of horrifying events a comet could be presaging. Adding to the amazing conclusions of Pliny was his list of the 10 types of cometary visitors. They were:

1. *Pogonias*, a comet with a beard or mane of hair hanging down.
2. *Acontias*, a comet vibrating like a javelin with a fast motion.
3. *Xiphias*, a comet with a short, pointy form like a dagger.
4. *Disceus*, a disk-shaped comet.

Figure 5.2. Comet Broughton (C/2006 OF2) showed a wedge-shaped tail on January 12, 2009, shot with an 8-inch Schmidt-Cassegrain scope at f/6.3, a CCD camera, and stacked exposures. Credit: Craig and Tammy Temple.

5. *Pitheus*, a comet shaped like a cask, and emitting a hazy light.
6. *Ceratias*, a comet with the appearance of a horn.
7. *Lampadias*, a comet appearing like a flaming torch.
8. *Hippeus*, a comet shaped like a horse's mane being blown by air.
9. *Argenteus*, a silver-colored comet of great luminosity.
10. *Hircus*, a comet influenced by goats and characterized by goatlike tufts of hair.

Pliny the Elder was not a scientist but rather a collector of information, good, bad, and indifferent. His hodgepodge of cometary facts included some glimpses of interest and truth but also, even in the context of the times, a large helping of utter nonsense. Others desperately needed to come along to help refine a more realistic view of comets as human beings saw them.

And that takes us back to one Claudius Ptolemy. Perhaps the last of the great free-thinking astronomers of the ancient world, Ptolemy was steeped in the intellectual vibrancy of Alexandria. Ptolemy is still best known for his masterwork, the *Almagest*, a complex book containing his views on the apparent motions of stars and planets. The text was completed as early as 150. Ptolemy's hypothesis of epicycles changed thinking about bodies' moving around in the cosmos. He believed that planets and other bodies move in epicycles, small circles, the centers of which form a larger circular motion around a body like the Sun. So as a smaller body orbits a larger one, it makes little circular loops as it moves around the larger one in a generally circular way.

The idea worked ingeniously, so accurately that it lasted until Johannes Kepler arrived with his more sophisticated (and accurate) ellipses in the 17th century. Because of this apparent leap forward, the *Almagest* was incredibly influential for centuries. And Ptolemy also penned the *Tetrabiblios*, an accompanying work on astrological themes – the earthly effects of astronomical bodies. Ptolemy left comets for the companion volume, considering them astrological bodies.

Thus, accomplished as his orbital mechanics were, Ptolemy considered comets signs of doom, wars, deaths of leaders, or other nonsense. In fact, he took the notions so seriously that he specified examples that could be used as a kind of guidebook. A comet's shape, he wrote, determined the nature of the impending horrific event and who would be stricken. The area of sky where the comet appeared and the tail's perceived direction dictated where on Earth the doomsday scenario would play out. If the comet appeared suddenly near the Sun in the sky, then the terror might happen quickly. If the comet appeared away from the Sun, more time might tick away before the ill effects struck. And the amount of time the comet lingered in the heavens could determine how long the terrible events would go on down here on Earth.

A work that followed and was reprinted many times throughout the Middle Ages, the *Centiloquium*, was erroneously attributed to Ptolemy. The work contains a hundred aphorisms about astrology and was first seen widely in the 10th century, a version containing a commentary by Egyptian mathematician Ahmed ibn Yusef (835–912). Among its other astrological rules and regulations, the work contains specific rules for comet-related disasters:

1. The appearance of a comet at a cardinal point 11 astronomical signs from the Sun implied that a kingdom's sovereign, or perhaps an associate or prince, would soon die.
2. The appearance of a comet in a succeeding house signaled prosperity for a kingdom's treasury but a change in the ruling monarch.
3. The appearance of a comet in a cadent house, one that had passed the meridian, meant disease and death.
4. If a comet moved from the west to the east, a country or region would be invaded by an enemy.
5. A stationary comet meant the enemy would emerge from within – from the country itself.

With the dissolution of Greek and Roman dominance and the entry of the world into that thrilling period known as the Middle Ages, progress on revising views on comets was slow. Dominated by the church, freethinking stifled by dogmatically looking backward instead of forward, the medieval period delivered a gloomy forecast for progress in any of the sciences.

One of the earliest medieval astronomers was the English monk Bede the Venerable (*ca.* 673–735), who lived and worked at two monasteries in modern-day Jarrow, Northumbria. This intellectual for his time believed that comets signaled wars, disease, change in rulers, heat waves, or fearsome winds. Like his predecessor Ptolemy, Bede laid down hard and fast rules about comets and their behavior. He claimed that some moved like the planets while others did not move at all. They nearly always appeared in the northern sky and usually were placed near the Milky Way. He repeated the exact periods of visibility given so many years before by Ptolemy.

Others joined the fray. The German Dominican friar and bishop Albertus Magnus (*ca.* 1193–1280 or 1206–1280) believed that comets were terrestrial vapors that emanated from lower spheres of the terrestrial region to the boundary of the fiery sphere, where they became visible. The comet's visibility was dependent on the amount of fuel it contained, and when that burned out, the comet disappeared. Magnus believed comets were symbols of terrible or important events and not their causes.

One of Magnus's students was none other than English philosopher Roger Bacon (*ca.* 1214–1294). Although in this period we still hadn't reached the scientific, observational rigor of Galileo, Bacon pushed toward systematic observations and experiments on which to base conclusions. In this, he practically stood alone in an age of darkness in which a blind adherence to conservative ideals and a lack of independent thoughts were the overarching norm.

In July 1264 Bacon observed C/1264 N1, a bright comet visible in Leo, Cancer, and Gemini. He picked up on the comet's stay in Cancer and how it was moving toward Mars in the sky. Because of Mars's nature as the god of war, Bacon believed the comet signified a coming war and general chaos in England, Spain, Italy, and other countries. A medieval hanger-on, although he lived more than two centuries later, was German monk and seminal figure of the Protestant Reformation Martin Luther (1483–1546). Luther went so far as to write that "the heathen write that the comet may arise from natural causes, but God creates not one that does not foretoken a sure calamity." He called comets "harlot stars" and "works of the devil."

In 1578 the Lutheran bishop of Magdeburg, Germany, Andreas Celichius wrote a passage in *The Theological Reminder of the New Comet*, which summarized much thinking about comets during the latter period of the Middle Ages. Of comets, humans, and gods, he wrote, "The thick smoke of human sins, rising every day, every hour, every moment full of stench and horror, before the face of God, and becoming gradually so thick as to form a comet, with curled and plaited tresses, which at least is kindled by the hot and fiery anger of the Supreme Heavenly Judge."

At the time, not everyone believed that comets were caused by mortal sins. But Western culture produced staggeringly little progress in thinking about comets over centuries. That situation would only give way to the first glimmers of slight change around the year 1264.

Flashes of intellectual progress began to emerge with the French Dominican philosopher Aegidius of Lessines (also called Giles of Lessines, *ca.* 1230–1304), a student of Thomas Aquinas. In 1264, Aegidius observed C/1264 N1 and began to think logically about what he was seeing. The comet was visible in the evening sky after sunset but then crossed over into the morning sky. Before this time, observers believed that comets seen on two different sides of the Sun – even separated by small time intervals – were separate physical objects. Aegidius believed he saw the same object simply in motion.

Then along came the Italian astronomer and mathematician Paolo dal Pozzo Toscanelli (1397–1482). In the 15th century, this intrepid observer began to expand thinking on comets by carefully observing six comets in 1433, 1449/50, 1456, 1457, and 1472. The comet he observed in 1456 was an early apparition of Halley's Comet. But the insight produced by Toscanelli would be lost to history for four centuries: His manuscript notebooks went undiscovered, sitting in the Italian National Library, until they were found in 1864. Among his papers were carefully drawn star charts on which Toscanelli traced the positions of comets relative to stars, perhaps the first to use careful measurements of distances and angles.

In the same century, Austrian astronomer and mathematician Georg von Peuerbach (1423–1461) observed the 1456 apparition of what would turn out to be Halley's Comet. He became the first astronomer to attempt measuring the distance to a comet using parallax, viewing the comet simultaneously from different locations on Earth and, by calculating the angles and motion, deriving a distance. At the time, the young scientist (who would die well before his time) was a teacher at the University of Vienna. He studied planetary motions and believed, as Aristotle had, that comets were hot exhalations from Earth. His parallax measurement concluded the comet was more than 1,000 German miles above Earth's surface.

One of Peuerbach's students was a young fellow named Johannes Müller of Königsberg (1436–1476), also known as Regiomontanus, a German mathematician and astronomer. Regiomantanus popularized the methods pioneered by Peuerbach for determining cometary distances by parallax. Although he died early as did his teacher – and may have been poisoned – his works pushed forward the concepts of analyzing comets by observation and mathematics.

As the 16th century dawned, other Europeans cooked up novel ways in which to consider comets. Italian mathematician, astrologer, and physician Gerolamo Cardano (1501–1576) took on comets as one of his many areas of research. He investigated the distances of comets, studying C/1532 R1, but rather than using parallax measurements, he simply noted the comet must be distant because it was traveling more slowly than the Moon. Late in life, he was imprisoned analogously to Galileo, but for him it was for having cast a horoscope for Jesus of Nazareth and claiming that his life's highlights were predetermined by the movements of stars.

Cardano wrote about comets in works titled *Of Subtlety* (in Latin, *De Subtilitate*) in 1552 and *Of a Variety* (in Latin, *De Rerum Varietate*) in 1559. He believed that comets

Figure 5.3. When this image was made on October 8, 2010, Comet 103P/Hartley 2 was brushing past the spectacular Double Cluster in Perseus, two open star clusters — NGC 884 on the left and NGC 869. The imager used an 8-inch f/4 astrograph and composited exposures. Credit: Chris Schur.

were like globes excited into glowing from sunlight, and that - like lenses - the tails shone from sunlight passing through the comet, which focused the sunlight and created a tail behind the nucleus. Comets appeared during periods when the air was dry, and they signified corruption, famine, and death.

One of the greats of the entire history of astronomy, Polish astronomer Nicolaus Copernicus (1473–1543), originator of the idea of the heliocentric universe, briefly treated comets in some of his writings. In his important 1543 treatise *On the Revolutions of the Heavenly Spheres* (Latin, *De revolutionibus orbium coelestium*), he mentioned comets, assuming they were terrestrial objects and believing their motions reflected those of the "daily rotation" of the spheres.

"It is said," he wrote, "that the highest region of the air follows the celestial motion. This is demonstrated by those stars that suddenly appear - I mean those stars that the Greeks called cometae or poganiae. The highest region is considered their place of generation, and just like other stars they also rise and set. We can say this part of the air is deprived of the terrestrial motion because of its great distance from the Earth."

Soon after the time of Copernicus, the Great Comet of 1577 (C/1577 V1) would have a profound effect on cometary investigations. As we've seen, this comet would be the first such object known to exist outside Earth's atmosphere. The views of Aristotle, which had stood the test of nearly two millennia, were shattered. But it was significant for other reasons too.

The comet was enormously bright, rivaling Venus. A number of astronomers in a variety of places made measurements of the Great Comet of 1577. The first to make his results known, however, was the German astronomer Michael Mästlin (1550–1631). Where others had failed, Mästlin succeeded by simple cleverness. He employed a thread to observe the comet and align its position with two neighboring stars, keeping them all on the same great circle. He then repeated the process with other sets of stars. Finally, he referred to the voluminous star catalog in Copernicus's masterwork, noting the stars' positions and connecting the pairs of stars.

He repeated the process every few hours. In this way, Mästlin noted movement of the comet relative to the stars. He observed the comet in this way from late November 1577 through early January 1578 and developed a hypothesis to explain the comet's motion. His idea was that the comet moved in a circular path outside the orbit of Venus. He invoked the old idea of epicycles to explain small differences in observations from theory. Although the comet's orbit was parabolic and he held a belief in its circular form, his idea could, for the first time, more or less explain the comet's motion.

The Danish nobleman and astronomer Tycho Brahe, whom we have met before, was obsessed with the Great Comet of 1577. Tycho had famously observed the supernova of 1572, and just two years before the comet's appearance, Tycho had established his handsomely funded observatories, Uraniborg and Stjerneborg, on the island of Hven (now called Ven). Incredibly richly funded, Tycho was in an astronomer's paradise. On the evening of November 13, 1577, he was at a pond catching fish for dinner when he suddenly spied a comet in the sky. He observed the comet furiously for the next 2½ months.

From his measurements, Tycho concluded the comet had to be at least 250 Earth radii away from Earth's surface. He clearly believed it to be more distant than the Moon (his figure was 52 Earth radii for the lunar distance). He recorded observational details of the comet faithfully, calling the head "whitish or Saturn-like" in color and describing the tail as a reddish dark color similar to a flame when viewed through a cloud of smoke. For all of his scientific descriptions and calculations, Tycho still attributed future events on Earth to this celestial visitor. "Although this comet appeared in the west and will realize its greatest significance in those lands that lie toward the west," he wrote, "yet it will also spew its venom over those lands that lie eastward in the north, for its tail swept thence."

Tycho's significant work on the Great Comet of 1577 was published in 1588 with the short title *Of the World* (Latin, *De mundi*). Tycho published his parallax determinations for the comet in this work, and one of the barriers to understanding comets that had stood for what seemed like unending centuries was finally removed.

The view of how comets were perceived from antiquity up through the Middle Ages would be incomplete if totally centered on Europe, however. Asian astronomers observed the sky extensively as long ago as did the Europeans. China, in particular, deserves a reputation of being a hotbed of skilled observers who made frequent and analytical observations of comets among other sky phenomena. And unlike in the academic European tradition, Chinese astronomers tended to be philosophers who were supported openly – and living in the quarters with – their rulers.

Even in antiquity, when Europeans were more oriented to theoretical conceptualizing – great energy put into great thinking – Chinese astronomers were more focused on the observing itself. The practice of Chinese astronomy was more centered on collecting observations than on preemptively hypothesizing explanations; as a result the Chinese did not have the dominating ideas of circular orbits and cycles and spheres that the early Greek and Roman astronomers had.

Chinese observers focused on a celestial coordinate system of rotation around the poles very similar to the modern celestial sphere. They were not limited in scope by the Sun's apparent motion across the sky – as the Greeks were – or by the overarching concepts of altitude and azimuth – as Arabic astronomers were. The Chinese divided the heavens into asterisms, not in the modern sense but equivalently to today's constellations. In the 3d century, Chinese astronomers recorded 283 asterisms in the sky – more than three times the number of the modern constellations.

During the excavation of a large Han tomb at the Mawangdui site in Changsha, China, in 1973, archaeologists found a manuscript of silk pages with voluminous notes about nature and containing 250 drawings. The book, titled *Divination by Astrological and Meteorological Phenomena*, dates to about 168 B.C. and the composition of the work has been traced to around 200 B.C. Twenty-nine of the book's drawings represent comets ("broom stars") and form the earliest surviving human representations of comets known. For the 29 different forms or shapes of comets they recorded, the authors provided corresponding omens for each. These included the perishing of a state, a years-long war, disease in the world, a revolt in the army, raising of arms and famine, a bumper harvest but accompanied by war, a calamity in the state, a 5-day rebellion, death among generals, and three small battles along with seven larger ones. Clearly, a predilection for warfare and a fear of political instability were dominant themes.

Observationally, Chinese astronomers were sharp. They became the first to notice and record the fact that cometary tails point away from the Sun, and they may have recorded observations of antitails.

In the year 635, Chinese astronomer and mathematician Li Chunfeng (602–670) penned the astronomical portion of a voluminous history called the *Book of Jin*, the official historical text treating the Jin dynasty, which lasted from 265 to 420. The work was published in 648. In the book, Chunfeng wrote about comets.

"Among ominous stars the first are the hui-xing," he wrote,

> commonly known as broom stars. The body is a sort of star while the tail resembles a broom. Small comets measure several inches in length, but the larger ones may extend across the entire heavens. The appear-

> ance of a comet predicts military activities and great floods. Brooms govern the sweeping away of old things and the assimilation of the new. A comet can appear in any one of the five colors, depending on the essence of that one of the five elements which has given birth to it.

"According to the official astronomers," Chunfeng continued, "the body of a comet itself is non-luminous but derives its light from the Sun, so that when it appears in the evening, it points toward the east while in the morning, it points toward the west. If it is south or north of the Sun, its tail always points following the same direction as the light of the Sun – then suddenly it fades. The length of the rays is a measure of the calamity foretold by the comet."

Later official histories of subsequent Chinese dynasties introduced additional cometary forms and their associated omens. Observationally, the term Chang-keng seems to have referred to a comet with two "horns" or tails, presumably on both sides of the nucleus, and therefore may refer to an antitail. These forms of comets were reported by Chinese observers who spied comets in 467 and 886.

As the world inched its way out of the dogmatic Middle Ages and began to see the Renaissance and then the Age of Enlightenment, work on astronomy and on comets in particular flourished, along with many other sciences. When the last great naked-eye astronomer, Tycho Brahe, died in 1601, that event marked not only the onset of telescopic astronomy but also the passing of Tycho's torch to his assistant. At the end of Tycho's life, in Prague, German mathematician and astronomer Johannes Kepler (1571–1630) served as Tycho's right hand.

Kepler is best known for his three laws of planetary motion, which describe the motions of planets and other bodies around the Sun. Kepler correctly described how the orbits of such bodies are ellipses with the Sun at the center of one of the foci, how a line joining the planet and the Sun sweeps out equal areas during equal intervals of time, and how the square of the orbital period of a planet is directly proportional to the cube of the semimajor axis of its orbit. Published in 1609 (the first two laws) and 1619 (the third), this systematic thought from Kepler liberated the astronomy world from the cumbersome and incorrect notions of circles, spheres, and epicycles that had persisted from the days of Aristotle.

Kepler was not only a theoretician, however, but also an energetic observer. His thinking on comets would evolve from a very early age. Many years after the fact, he wrote a letter describing how when he was just 6 years old, his mother had taken him outside to view the Great Comet of 1577. Oddly, none other than the German astronomer Michael Mästlin, who had also observed the 1577 comet, later became Kepler's teacher at the University of Tübingen. By 1602, when Kepler was 31 and several years past publishing his first major astronomical work, *The Cosmographic Mystery* (Latin: *Mysterium Cosmographicum*), he had come to believe that comets were ephemeral, rocketlike objects with straight-line paths.

In his 1604 work *The Optical Part of Astronomy* (Latin: *Astronomiae pars optica*), Kepler wrote about comets. Strangely, although he had been taught by Mästlin, who believed in circular orbits for comets, Kepler never wavered in his peculiar belief in their straight-line paths. He could not accept comets as objects that warranted permanent inclusion in the celestial sphere – they had to be temporary nuisances. Kepler even outlined a careful experiment he had conducted in a dark room, shining a light onto a water-filled balloon or globe, which produced a reflected beam of light similar to a comet's tail. He believed the light passing through the head was deflected somehow toward an observer, creating the tail effect.

Kepler's next work on comets was a popular writing he produced in the wake of observing, for several weeks, the comet of 1607, which turned out to be Halley's Comet. He viewed the comet on each possible night in Prague while standing on a bridge spanning the Moldau River. He published the observations in 1619 in the work *Three Lists of Comets* (Latin: *De Cometis Libelli Tres*). Kepler surmised that comets came about spontaneously, from impurities or fatty globules in the ether. He believed that a special spirit guided comets through space and that space was filled with innumerable comets – which could only be seen when they approached very close to Earth.

The great scientist also offered further explanations for the formation of cometary tails. He now proposed that sunlight entered a comet's head, pushing a part of the head away from the Sun, which drew it out and created a tail. This began to dissipate the comet and the comet faded away and was gone when the material ran out in this way. (But this was not equivalent to saying that radiation pressure pushed particles away from the Sun.) The nonlinear forms of a comet's tail could be explained by refraction – not because the tail was actually curved. He also stated that should Earth have contact with a portion of the tail, the results for living beings on Earth could be disastrous.

Simultaneous with Kepler's cometary adventures in Prague, Italy hosted another clever fellow, physicist, astronomer, and mathematician Galileo Galilei (1564–1642). Galileo's career is so well known and his role as the first full observational empiricist so familiar that it doesn't bear repeating here. Suffice it to say that when he heard about the commonly available hand telescopes invented by Dutch lens makers that were for sale on the streets of Paris, he was horrified. Driven by the need for money, ambition, and knowledge, Galileo treasured the concept of a telescope because he knew that it would be highly valued by the doge in Venice, whose approval (and funds) he sought. Rushing to his workshop in Padua, he reinvented the telescope simply on the basis of what he had heard of it, and by the fall of 1609 he was testing it.

One evening Galileo climbed to his rooftop, moved the field of view of his simple 3× telescope (with about a 1-inch aperture) from the church steeple near his house over to the Moon and made the first-ever influential telescopic observation of an astronomical body. A new era of science was born. Lauded as one of the great figures of science, by some – Carl Sagan included – Galileo has outright been called the father of science.

When three bright comets appeared in 1618 – C/1618 Q1, C/1618 V1, and C/1618 W1 – the observers of the world took note and watched them carefully. Such was not the case with Galileo, however, who at the time was bedridden with arthritis and a double hernia. Not so with Italian Jesuit mathematician and astronomer Orazio Grassi (1583–1654), who observed the comets and in 1619 anonymously wrote *Three Comets of the Year 1618* (Latin: *De tribus cometis annus MDCXVIII*). Grassi looked forward in this work, abandoning the ordinary and astrological fears of comets and supporting Tycho's stand on cometary explanations. He supported comets as distant objects on the basis of parallax measurements, suggested they moved in great circular orbits, and imagined they shone by reflected sunlight. They were travelers of greater distance than the Moon and less than the Sun, indicated by their intermediate speeds in the heavens.

Galileo, being bedridden, participated in the great discussions over the three bright comets by enlisting his student Mario Guiducci (1585–1646), who penned a detailed treatise on the comets, *Discourse on Comets* (Latin: *Discorse delle comete*), in 1619. Although it appeared in Guiducci's name, it was essentially the work of Galileo. Rather than presenting his own ideas, however, Galileo took up the opportunity to attack the views of Grassi.

Galileo took sharp aim. He suggested that comets were not periodic. He described how the only other bright comet that would rival the brightest comet in 1618 was the comet of 1577, and those two could not be the same object. This comet would not have traveled a single degree over the 40 years, Galileo underscored. Galileo also questioned whether comets were "real" rather than reflections. How could Grassi know that comets were more distant than the Moon if he didn't know whether they were real or reflections? With his own telescopic observations as a basis, Galileo suggested that any enlargement of an object in a telescope is independent of its distance, and therefore Grassi's suggestion that comets were distant because of their telescope appearance was flawed.

Galileo also suggested that comets may form close to Earth's surface and move upward slowly, causing observers on Earth to see them as large but without a tail at first, may glow by sunlight reflecting off a "cloud of vapors," and show a curved tail due to refraction or simple perspective effects. He also attacked Tycho on the issue that cometary tails appear to point away from Venus, blasting the dead Dane for connecting the two when Venus's light could hardly illuminate such a wispy body.

Grassi replied to Galileo's criticisms by issuing a text titled *Astronomical Scale* (Latin: *Libra astronomica*), in 1619, under the fanciful pseudonym Lothario Sarsi. Grassi attacked Galileo for associating comets with Earth (which he had not straightforwardly done), criticizing the astronomer with all the now-familiar arguments against comets as products of Earth: If issued as "exhalations" from Earth and rising upward, they would be swept away by winds, would reflect the Sun's light as rainbows do, and so on. Grassi then suggested that comets could move in elliptical orbits, although he did not believe in a Sun-centered cosmos.

Controversies come and go, but this one stayed around. The next move was that Guiducci sent a letter of rebuttal to the Jesuit scholar Torquino Galluzzi (1573–1649), a letter he also openly published. Galileo fired the next volley in 1623 by publishing *The Assayer* (Italian: *Il saggiatore*). Before addressing the present controversy, Galileo used the diatribe to attack German astronomer Simon Marius (1573–1624), who had claimed to see the four Galilean moons of Jupiter before Galileo did. After correcting him on that count, Galileo then claimed he had not denied the possibility that comets were more distant than the Moon. He had simply chosen a comet moving the way it did because it best suited his point.

"Tycho himself," Galileo wrote, "among so many disparities, chose those observations which best served his predetermined decision to assign the comet a place between the Sun and Venus, as if these were the most reliable."

And then Kepler entered the fray. Having seen his master attacked, he responded in a work designed to defend Tycho against several other critics, *Shieldbearer of Tycho Brahe, the Dane* (Latin: *Tychonis Brahei Dani hyperaspistes*), published in 1625. An appendix created for this work addressed the controversies that arose from the comets of 1618. It also stepped aside on the issue of the glass-globe-and-reflections experiment Kepler had conducted in 1604. He proposed revised views on the formation of cometary tails.

"The head is like a conglobulate nebula and somewhat transparent," wrote Kepler; "the train or beard is effluvium from the head, expelled through the rays of the Sun into the opposed zone and in its continued effusion the head is finally exhausted and consumed so that the tail represents the death of the head."

Kepler also argued against Galileo's views on the tail curvature issue. If curved comet tails were caused by refraction of light, Kepler argued, then the curvature would be slight and would always point upward toward the zenith – and only when comets were near the horizon. Kepler also stuck up for his old master Tycho, describing the meticulous nature of his work. Addressing the Jesuit's refusal to adopt the Copernican universe, Kepler shot back that their notions were "perverse and querulous at best, servile at worst."

By the mid-1620s, Grassi's arguments were running out of steam. He fell so low as even to suggest that Galileo's argumentative nature resulted from drinking too much wine and "joked" that *saggiatore* translated into "winetaster" – *assaggiatore*. At long last, the raging debates ground to a halt.

For some 40 years the cultural standing of comets stayed on steady footing. In November 1664 Spanish observers found a bright comet 18 days before it passed

Figure 5.4. This mosaic captures the fine details of Comet NEAT (C/2002 V1) as they appeared February 7, 2003, recorded with an 8-inch Schmidt-Cassegrain scope at f/2.8, a CCD camera, and stacked exposures. Credit: Dennis Persyk.

perihelion. By year's end it would pass close to Earth (25.4 million km) and unleash a major round of observations. The Comet of 1664 would later be designated C/1664 W1.

At its brightest, C/1664 W1 reached magnitude 0 and sported a tail some 40° long. Another, dimmer comet appeared the following year, C/1665 F1. The appearances of these two comets, at their brightest just 4 months apart, set astronomers to work and produced publications that reexamined cometary science.

The preeminent observer and theorist of the moment was Italian-French astronomer and mathematician Giovanni Domenico Cassini (1625–1712), best remembered as the namesake of the dark Cassini Division in the rings of Saturn. Cassini spent much of his career at Panzano Observatory, a private facility at Bologna, built by a wealthy astronomy enthusiast. In 1669 he moved to Paris and a short time later became director of the Paris Observatory. (In France, he became known as Jean-Dominique Cassini.)

Using his observations of C/1652 Y1, Cassini suggested that comets were beyond the orbit of Saturn. He revised this, however, to conclude that comets are similar in their distances and orbits to planets. With his observations of C/1664 W1, Cassini produced a mountain of new work. He recorded his new concepts in *Last Hypothesis on the Movements of Comets* (Latin: *Hypothesis motus cometae novissimi*), in 1665.

Cassini invoked the orbital systems he was familiar with – Earth and the Moon, and Jupiter and its four Galilean moons. Cassini then worked observationally and mathematically to find the central body the comet appeared to be orbiting. When the comet was closest to Earth, he calculated the comet must be orbiting Sirius, the brightest star in the sky. He suggested the comet was orbiting Sirius in epicycles and Sirius, in turn, was orbiting Earth.

When the Great Comet of 1680, Kirch's Comet, arrived, Cassini had a go with determining its orbital motions. Most still believed the comet observed before and after perihelion represented two objects, and Cassini struggled with making the observations fit a circular orbit. He suggested that comets could be restricted to a zodiaclike band similar to that of the planets.

Another prominent observer and theoretician focused on these comets was the French astronomer Adrien Auzout (1622–1691). Born and educated in Rouen, he left for Paris in the 1640s and became an energetic observer well before the appearance of C/1664 W1. After his initial observations, Auzout produced an ephemeris that accurately predicted the upcoming positions of the comet. Auzout plotted the comet's observed positions on a large celestial globe and then angled the globe to produce a horizon circle, enabling the plotting of future positions. The technique worked exceptionally well and was also employed by Cassini. Auzout agreed with Cassini on the issue of C/1664 W1 moving in a circular orbit about Sirius.

At about the same time, Louis XIV of France (1638–1715), the Sun King, asked for scientific research to be done on comets, fed up with superstitious fallout from comets and how it might affect his young rule. He sought Pierre Petit (1597–1677), the royal geographer, to write *Essay on the Nature of Comets* (French: *Dissertation sur la Nature des Comètes*), published in 1665. Petit was a rationalist, and a good deal of the work served as a public relations campaign against irrational fears about comets. Petit wrote about comets as permanent bodies in large orbits that occasionally made their way in to the vicinity of Earth and believed they performed a kind of cleaning of space, mopping up dangerous gases and other poisons emitted by planets and other bodies.

To address the issue of how far away comets could travel, Petit turned back to the argument used by Grassi against Galileo. He concluded that comets were very distant because they could not be magnified in a telescope as effectively as planets could be. Unsuccessful with parallax measurements, Petit reasoned mathematically, suggesting that comets could have periods as long as 1,000 years or more. He also believed that comets probably traveled in elliptical orbits but would not specify that the center of the orbits was either Earth or the Sun.

A geographer studying comets who had a rich history of astronomers and mathematicians preceding him, Petit was an unlikely person to be the first to predict the return of a comet. But that's exactly what he did. Petit took notice that C/1664 W1 and one of the comets visible in 1618 were probably one and the same object, on the basis of apparent orbital and observational similarities. He then traced other possible comets backward from 1618 at 46-year intervals. He also predicted the comet's return in 1710, taking immense pride in becoming the first person to predict the return of a comet. But Petit was wrong: The two comets were not the same, and in 1710 the comet didn't come back.

Further observations of C/1664 W1 led in other directions. Italian physiologist, physicist, and mathematician Giovanni Borelli (1608–1679), a believer in Galileo's method of repeated observations and testing, observed the comet for 3 months. He measured the comet via parallax and determined that it was more distant than the Moon. He based his calculations on the comet's orbit on the assumption that the Sun was in the center of the universe and found that the comet traveled in what was probably an elliptical path and that its distance from Earth varied over time. Because of the church's anti-Copernican bias, Borelli summarized his conclusions gingerly and published his results under a pseudonym.

It was this same comet, C/1664 W1, that also produced the first systematic observations of a comet from a North American observer. Samuel Danforth (1626–1674) of Roxbury, Massachusetts, was a man of many trades – Puritan minister, preacher, poet, and astronomer. After his 3 months of observations, Danforth published a treatise on the comet titled *An Astronomical Description of the Late Comet or Blazing Star*, in 1665. A careful thinker, Danforth concluded the comet was traveling in a circular orbit offset from the center of Earth. Because the comet moved slowly, he assumed it was more distant than the Moon. He measured the comet's motion and suggested that because it moved fastest on December 28, 1664, it passed closest to Earth on that day (perigee was on December 29). He believed sunlight passed into the transparency of the comet, refracted through it, and became visible as a stream behind the comet's head.

One of the great scientists of the era, English philosopher and polymath Robert Hooke (1635–1703), also observed and spent time thinking about the comets visible in 1664 and 1665. He also did some background research by extensively analyzing observations of the Great Comet of 1577. He waited to publish any results, however, hoping to observe another brilliant comet that would allow him to test some of his hypotheses. So when a bright comet came along in 1677 (C/1677 H1), he viewed it carefully, conducted experiments, and finally readied to publish his findings.

His work, published in 1678, was titled *Cometa or Remarks about Comets*. In it, he summarized his careful telescopic observations, made with a 6-foot telescope, which he had set up such that he could compare the size of the comet's coma carefully with that of a nearby pole that supported a weathervane. His conclusions were many. He believed that comets were solid bodies made of magnetic materials similar to those residing in Earth's interior. He suggested that the nucleus could dissolve chemically from the ether that surrounded it and suggested that – like magnetite, or lodestone as he called it – a comet could become repulsive, "confounding" gravity and allowing material to stream off the nucleus, forming a tail. Matter would fly off the nucleus in all directions but be blown back away from the Sun because of the material's repulsive nature.

Hooke also noted that because shadows were not visible on a comet's nucleus, it was possible the comet produced some light of its own. He invoked other bodies that produce light of their own – the Sun, stars, or even light from decaying garbage. Hooke also suggested that a comet's magnetic "virtue" could wind down over time, pushing the comet's orbit from an arc closer to a straight line. He also mentioned the role of gravitation in the lives of comets: "I cannot imagine how their various motions can with any satisfaction be imagined," he wrote, "without supposing a kind of gravitation throughout the whole Vortice or Coelum of the Sun, by which the Planets are attracted, or have a tendency toward the Sun, as terrestrial bodies have toward the center of the Earth."

At this point, the man who is often regarded as the greatest scientist in history enters the picture. English physicist and mathematician Isaac Newton (1642–1727), Lucasian Professor of Mathematics at Cambridge University, received a letter from his colleague Hooke. In it, Hooke stated his belief that gravity's effect decreased proportionally with the square of the distance between two bodies. The exchange of letters that followed led Hooke to believe, later on, that Newton had taken the idea for his famous inverse square law of gravitation from him. Newton countered that Hooke's own work on comets did not include any such discussion – and that even English architect Christopher Wren (1632–1723) knew of the gravitational relationship before Hooke.

This controversy aside, Hooke concluded his work on C/1664 W1 by announcing that the comet showed no parallax effect, as simultaneous observations had been inaccurate. He did know the comet was distant. He, like Newton, firmly believed in the Copernican universe, and beyond that, he could conclude little about the precise nature of the comet's orbit.

Polish councilor, mayor, and astronomer Johannes Hevelius (1611–1687) also exuded a great influence on cometary science during this period. Although he was born into a wealthy family of brewers in Danzig and received financial support from Louis XIV as a patron, Hevelius would later fall into controversy, a tragic fire, and ill health.

The first of the controversies rose over the positions Hevelius published for C/1664 W1. Auzout, Petit, and others attacked him vigorously over using naked-eye observations for determining the comet's place in the sky. Hevelius observed comets telescopically but then made positional measurements with his eyes alone, not trusting the transfer between a telescopic field and his view of the sky. The controversy flared to a much greater extent when, subsequently, Hooke and England's first Astronomer Royal, John Flamsteed (1646–1719), entered the fray, also attacking Hevelius.

During this period of controversy in 1679, Hevelius was struck by tragedy when his observatory in Danzig, Stellaburgum, burned to the ground. He had only partial success in rebuilding the structure and repopulating it with telescopes that were only a shadow of his former assemblage of equipment. The loss of his equipment and reputation bit hard, his health declined, and Hevelius died on his 76th birthday.

Before the full force of tragedy, however, at the end of 1664, Hevelius was already deeply involved in writing his masterpiece on celestial visitors, *Cometographia*. Because of the great public interest in comets at the time, Hevelius produced a synopsis of the major work that was titled *Harbinger Comets* (Latin: *Prodromus cometicus*).

In the short work, Hevelius announced a parallax of 5,000 Earth radii – actually one-sixth of the real distance. He proclaimed that comets are spurious planets, as they orbit the Sun as planets do. He supposed that taking Earth's motion into account was important for determining the motions of comets, but he did not account for Earth's annual orbital motion, and so his calculations were flawed. He later corrected this discrepancy in the full work.

Hevelius stated that C/1664 W1 was very close to the star Alpha Arietis on February 18, 1665, a date that was important in his calculation of the comet's parallax. But other observers quickly pointed out that by their – consistent – measurements, the comet was 6.5° away from the star on that date. And when, three weeks later, the comet passed close to a bright star, it was Beta Arietis, not Alpha.

Rather than admitting the error, though, Hevelius launched into a long defense of his unsupportable position. The matter went on as a long series of arguments, and Hevelius only backed down after this protracted mess when the Royal Society in London refused to accept his observations. Still, much of what Hevelius published about comets in *Cometographia*, printed as a series of papers over 15 years and leaning generously on Kepler, held great value for the scientific world.

When it came to progress with understanding comets, few eras could match the meaning of Isaac Newton's renewed interest in the subject that accompanied the comet that would come to be known as C/1680 V1, the Great Comet of 1680, or Kirch's Comet. This, the first to be found telescopically, delivered a huge impact on the interpretation of comets. And, as usual, Isaac Newton was at the forefront of this progress.

Newton observed the comet, thought about it carefully, and calculated details of its physical nature. At first Newton believed comets seen in November and December 1680 were two different objects – on different sides of the Sun – moving on rectilinear paths. But after seeing correspondence about the comet between his colleagues Flamsteed and Edmond Halley, he began to ponder this comet seriously.

Figure 5.5. Comet 29P/Schwassmann-Wachmann (Schwassmann-Wachmann 1), a periodic comet that reaches perihelion once every 14.7 years, appears as soft as a cotton ball when this image was made March 18, 2010, using an 8-inch f/4 reflector and stacked exposures. Credit: Chris Schur.

On February 28, 1681, Newton wrote to Flamsteed that he still believed the November and December observations represented two objects. He offered some advice on what he believed was Flamsteed's erroneous conclusion that the observations were of the same comet, having moved to the other side of the Sun. He pointed out that magnetism could not be responsible for pulling the comet around the Sun because a red-hot lodestone (magnetite) loses its magnetism, and the Sun was certainly quite warm. Newton argued that if the observations represented the same comet, then it would have had to undergo rapid acceleration and deceleration that wouldn't make sense.

But Newton had been using the observations that were available (not all his own), and some were flawed. When he replied on March 7, 1681, Flamsteed corrected the erroneous observations and suggested the Sun's magnetism may not be like that of a lodestone. Newton held on, but began to convert to the one-comet hypothesis. He wondered why other, earlier comets weren't visible passing around on various sides of the Sun.

Three years later, Newton had come to accept the idea that comets travel on closed elliptical orbits. It was partly from the mystery of explaining cometary orbits that he set about producing his masterwork on gravity and allied subjects, *Mathematical Principles of Natural Philosophy* (Latin: *Philosophiae Naturalis Principia Mathematica*, or *Principia* for short), in 1687. None other than Edmond Halley underwrote the publication and acted as publisher.

In the last of the work's three substantial books, Newton outlined his method for determining the parabolic orbits of comets. He employed three identical observations that were nearly evenly spaced in time. As an example, he provided the details for his analysis of the orbit of the Great Comet of 1680. The method was still somewhat crude relative to modern precision, of course, but now completely correct in spirit and analytical approach.

Thus, Isaac Newton had closed the loop on comets as objects of total mystery. A great deal of work was left until humans really began to understand comets. Fred Whipple was not quite born yet. But in 1,700 years, philosophers, astrologers, and astronomers had gone from pure superstition and speculation on the nature of comets to a mathematical understanding of them and their place among other celestial bodies. Seneca had wondered on paper whether humans would ever know whether comets were portents or stars.

It turns out that they are something else entirely.

6

Where Comets Live

The first step in understanding anything about a comet is to understand where it is. And the knowledge about where comets are, how they move, where they come from, and how they start in toward the inner solar system has changed dramatically over the years. Comets provide a good framework for understanding the distance scale of the solar system, as well as contemplating individual objects and why they behave the way they do.

When someone reports a suspected comet discovery, the first thing astronomers do is to make precise astrometric (positional) measurements of the object as it moves along its orbit. They want to make these accurate measurements of the comet's position over a reasonably long arc – that is, over three or more measurements over the span of several days or a week – in order to calculate a preliminary orbit for the object. To do this, they need more than just careful positional measurements. They also consider the comet's center of mass, gravitational perturbations that may be influencing the comet by planets or asteroids, and the modeling of nongravitational forces – that is, any outgassing from the comet itself that would influence its orbit.

To produce the astrometric data that astronomers need to calculate an orbit, they collect positional measurements as pairs of right ascension and declination coordinates – the equivalents of longitude and latitude on the celestial sphere. They offset the apparent position of the comet from stars of precisely known positions, and ideally these are from catalogs that contain star positions from the *Hipparcos* star catalog. The European Space Agency's *Hipparcos* satellite made the highest-accuracy positional measurements of stars to date during its lifetime of 1989–1993.

The best potential positions for comets occur when the object is closest to Earth, but with comets that also means the coma is at its largest, and the nucleus is correspondingly hard to see. So astronomers assume that the brightest part of the comet's light also represents the comet's center of mass (its nucleus) – but sometimes this

Figure 6.1. The comet with the shortest period known, 2P/Encke, orbits the Sun once every 3.3 years. This image shows it as a fuzzball and was taken on November 17, 2003, with a 4-inch refractor, a CCD camera, and stacked exposures. Credit: Mike Holloway.

isn't the case. With modern CCD observations, astronomers have found that the brightest individual pixel typically represents the comet's nucleus, not the brightest part of the overall cloud of light as determined mathematically.

The trickiest part of predicting a comet's future motion is reading the tea leaves from the nongravitational effects. No comet has underscored the unpredictability of these effects throughout history better than 2P/Encke, the periodic comet that has often shown bursts of rocketlike outgassing (Figure 6.1).

French astronomer Pierre Méchain (1744–1804) discovered Comet Encke in 1786, and the object was subsequently independently found by a Who's Who of comet observers including German-English astronomer Caroline Herschel (1750–1848, sister of William Herschel) and French astronomer Jean-Louis Pons (1761–1831), who became one of the greatest pure visual discoverers of comets ever, with at least 27 finds. It was German astronomer Johann Encke (1791–1865), however, who produced the numerical computations to show that these many comets discovered by all the others were in fact the same object that now bears his name.

In 1823, Encke wrote that he believed the comet moved against some medium, perhaps an extension of the Sun's atmosphere or debris from planets or comets that influenced the comet's motion in space. Understanding its behavior allowed him to predict the comet's return in 1825 and 1858 with accuracy. But the medium in space hypothesis didn't fare so well with German mathematician and astronomer Friedrich Wilhelm Bessel (1784–1846), who believed that a comet outgassing material in a radial way, toward the Sun, would also experience a recoil force that would shorten or lengthen the comet's orbit depending on the amount of material outgassed before or after perihelion. This was the genesis of the nongravitational force idea for comets, and this notion caught on and ultimately proved to be correct.

During the second half of the 19th century Comet Encke indeed behaved differently than Encke had predicted, and studies with other comets as late as 1940 showed the idea of a resisting medium in space was incorrect. The nongravitational force idea of Bessel's gained momentum, although no one yet knew that the forces were unleashed by ices warming and blasting out jets of water and gases vaporizing from the comet's nucleus.

In fact, planetary scientists had to wait until a century after Bessel's death to produce a sophisticated modeling of the nongravitational effects of comet nuclei, and this occurred with Fred Whipple's dirty snowball model in 1950. In fact, understanding the nongravitational forces and finding a solution to this puzzle were a large part of what drove Whipple to hypothesize the dirty snowball model in the first place. Astronomers considered the orbits of comets they knew well - periodic comets with relatively short periods.

After planetary scientists took into account the perturbations from planets, they still couldn't explain the accelerations and motions of many of these objects in a precise way without the nongravitational forces. They concluded that when ices sublimate from a comet's nucleus, the process transfers momentum to the nucleus, in effect giving it a little nudge. Over time, these effects produce deviations in a comet's time of perihelion passage, for example, as many as 4 days per apparition in the example of Comet Halley.

On the basis of studies of 1P/Halley by the *Giotto* spacecraft in 1986 and on studies of 19P/Borrelly by *Deep Space 1* in 2001, planetary scientists now believe that this outgassing on the surface of a comet's nucleus takes place at widely scattered, discrete areas where the comet happens to be more porous. They used to believe the outgassing took place preferentially on the comet's sunlight half, but the spacecraft analyses of Halley and Borrelly tended to refute that explanation.

From what planetary scientists have observed with Halley and in the years since, it appears that comets are very much individuals (Figure 6.2). Each comet displays its own set of jets and outgassing located at different areas, depending on the individual makeup of the comet's nucleus. Thus, a very detailed model of how a comet outgasses would require a study of each individual comet. So calculating orbits for comets that have not been studied up close with spacecraft will continue to require some assumptions in the model. But astronomers use sophisticated models that account for a comet's rotation, its oblateness, the positions of potential outgassing, and other factors.

Once they understand the basics of a comet's orbit, of how it's moving, astronomers can begin to place it in the context of the solar system as a whole. As we've seen, philosophers/mathematicians/astrologers/astronomers considered comets to

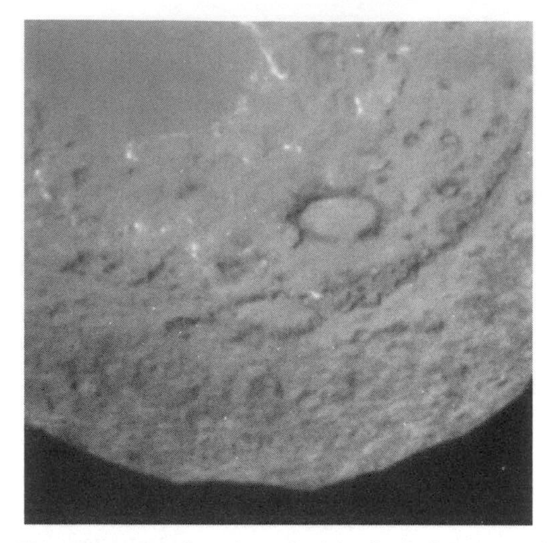

Figure 6.2. The cratered and pockmarked surface of Comet 9P/Tempel 1 appears grotesquely battered in this view from the Deep Impact spacecraft 90 seconds before it struck the comet on July 4, 2005. Credit: NASA/JPL-Caltech/UMD

be atmospheric phenomena during the majority of the last two millennia. Only in the last 400 years has it been generally accepted that comets are astronomical objects and exist at distances well beyond the Moon.

Amazingly, however, it took until the mid-20th century, the time of Fred Whipple, to clear the air of another outstanding puzzle. Only then did astronomers universally come to believe that comets are really members of the solar system rather than interlopers from interstellar space that just happen to be whizzing through our neighborhood. And pretty good reasons existed for this uncertainty, because the majority of comets do come from a very long way away.

Given the incredible uncertainty of most early observers over the distances to comets, it's impressive that by the mid-19th century the majority of scientists believed that comets have orbits larger than those of the planets. The modern division of cometary orbits, and therefore groups or families of comets, originated with Irish astronomer and science writer Dionysius Lardner (1793–1859), compiler of the 133-volume *Cabinet Cyclopedia*, who wrote in 1853, "we are in possession of the elements of the motions of 207 comets. It appears that 40 move in ellipses, 7 in hyperbolas, and 160 in parabolas."

Lardner's analysis suggested three classes with elliptical orbits, which have come to be known as Jupiter-family comets, Halley-type comets, and long-period comets. Hyperbolic and parabolic orbits are new long-period comets whose orbits have not yet evolved into huge ellipses. Lardner also looked at the direction of cometary motions, noting that – not counting the Jupiter-family comets – there are roughly equal numbers of comets that orbit in the same direction as the planets and those that orbit in the opposite direction.

The murky origins of comets received further study from American astronomer Hubert A. Newton (1830–1896) and Dutch astronomer Adrianus J. J. van Woerkom (1915–1991), who focused on the gravitational effects of Jupiter. In 1948 van Woerkom studied the orbits of comets and demonstrated that their energies were not consistent with an interstellar origin. Two years later, Dutch astronomer Jan H. Oort wrote his game-changing paper that tied the questions of cometary origins together. "There is no reasonable escape," he penned, "I believe, from the conclusion that the comets have always belonged to the solar system. They must then form a huge cloud, extending … to distances of at least 150,000 [astronomical units] and possibly still further."

So the Oort Cloud, the distant, enormous shell of comets surrounding the solar system, began to be proposed by astronomers to be the principal storehouse of comets in our vicinity. This is the home of long-period comets, from which these objects in their huge orbits flutter out to incredible distances and with orbits that remain stable over billions of years, sometimes traveling in close to our neck of the woods, warming up, and becoming gossamer parts of the sky. Just exactly why do astronomers believe this to be so? And how do long-period comets behave?

Along with the planets, the young solar system contained a large number of so-called planetesimals, small icy and rocky bodies between the planets and on the solar system's outer edges. The region of about 4 to 40 astronomical units from the Sun (from a little closer than Jupiter to the region just beyond Pluto), where ices would be able to be vaporized, would have contained countless planetesimals. Most of these planetesimals would have been perturbed by the planets into orbits that would have crossed the plane of planetary orbits, sweeping them up. This is except for the Kuiper Belt, the region of small, icy bodies concentrated between about 30 and 50 astronomical units out, where the orbits could remain stable for billions of years.

Long-period comets were probably born from the planetesimals that didn't remain in stable orbits. Planetary scientists think the movement of a comet from the outer area of planets all the way out into the Oort Cloud happens over several discrete steps. First, perturbations from planets need to kick the icy object into a more energetic orbit – that is, keeping its perihelion distance about the same while it lengthens the semimajor axis of the orbit. This makes the orbit larger and larger, sending it out deeper into the abyss at its farthest point from the Sun.

But there has to be a limit to this process, or the comet would eventually escape into deep space, captured by the gravity of a passing star or cloud of gas while the Sun loses its gravitational hold on the comet. Decades ago, astronomers argued that stars and other distant objects can actually provide a stabilizing effect on comets undergoing this lengthening of their orbits. In 1932 Estonian astronomer Ernst Öpik and, 18 years later Oort himself each suggested that stars could indeed affect the orbits of comets. But rather than stealing the comet by gravity, a passing star could place it neatly into the Oort Cloud by readjusting the comet's perihelion away from the region of the planets, before planets knocked the comet out of the solar system. Comets that stay in the Oort Cloud have their orbits perturbed into a large semimajor axis that is affected by passing stars but are not perturbed so much as to push it so far away that they are weakly bound to the solar system.

Planetary scientists thus believe that comets that were strongly perturbed by planets and placed into orbits with semimajor axes of 10,000 to 100,000 astronomical units inhabit the Oort Cloud. Comets perturbed out to greater distances probably escaped the Sun's gravity, and those not perturbed enough crossed the plane of the planets and were long ago accreted into larger bodies. The great shooting gallery of the solar system found a way to incorporate many small bodies into the larger ones, to sling most of them far away into a frozen holding bin, and to eject many others into interstellar space. But because passing stars do so in random directions, some of the Oort Cloud comets they knock about by gravity find their paths directed inward toward the Sun, reducing their perihelion distances and sending them toward us for a close encounter with the inner solar system.

But stars alone do not make a galaxy. The Milky Way Galaxy contains some 400 billion stars, perhaps (we don't know exactly how many dwarfs it holds), but also copious amounts of gas, dust, and dark matter. For some years astronomers have realized that the so-called galactic tide, consisting of a flow of gas clouds, dark matter, and so on, also influences comets. The majestic time scale of a billion years or more can be involved with the passing of the galactic tide, which sees cometary perihelion distances vary from longer to shorter in variable time frames. And sudden, unpredictable events can also influence comets: Giant molecular clouds can knock them free of their normal orbits on shorter time scales.

There are limits to these new interlopers from afar. What planetary scientists call dynamically new comets – those on freshly altered paths and brand-new visits toward the Sun – usually originate at or beyond tens of thousands of astronomical units from the Sun. "New" comets from this distance moving in close to the Sun from the Oort Cloud's inner edge must have had a previous perihelion distance greater than 10 astronomical units, about the distance of Saturn from the Sun, or they would have been either captured by Jupiter or Saturn or knocked far out of the solar system by their force.

Thus, astronomers know that such fresh comets from the inner Oort Cloud should only enter the inner solar system after a strong perturbation by a passing star, the push of a giant molecular cloud, or another unusual gravitational event.

Figure 6.3. Comet 8P/Tuttle has a period of 13.6 years and appeared as a large, diffuse halo when photographed on December 27, 2007, using a 20-inch Ritchey-Chretién scope at f/5.2, a CCD camera, and stacked exposures. Credit: Paulo Candy.

It was Oort himself who recognized that long-period comets must originate from a huge, spherical cloud at 10,000 or more astronomical units from the Sun. He suggested that most new comets reach their most distant points in their orbits at about 100,000 to 150,000 astronomical units. Recent work suggests figures that may be as small as half these values. And the issue of nongravitational forces muddies the water in trying to determine precise orbital values, because they make cometary orbits look more eccentric than they really are.

What happens to so-called long-period comets over time? In one recent study of 386 such comets, planetary scientists found that 25 percent of them are in slightly hyperbolic orbits, meaning they will never again return to our neighborhood and will exit the solar system unceremoniously. In their first appearances close to the Sun, roughly half of new comets are ejected by Jupiter and the other large planets into interstellar space. Most of the other half are captured into smaller, less eccentric orbits. Only 5 percent of fresh long-period comets return to Oort Cloud distances. On subsequent returns, many of these comets will be ejected, captured into a short-period orbit, or destroyed by collision with the Sun or a planet.

And there are other ways in which old comets can die. Like old soldiers, they can simply fade away. Oort noted that the number of comets that should undergo return visits was far less than would be expected. He called this phenomenon cometary fading. It does not refer to a fading in brightness, but to a variety of effects that may include random disruption or splitting from stresses or impacts, loss of volatiles, and/or the formation of a crust or mantle on the nuclear surface that would prevent outgassing. If these mechanisms or others turn a formerly active comet into an inactive comet, astronomers refer to the object as dormant, extinct, or disrupted.

To consider the number of Oort Cloud comets in existence, Oort turned to the number of fresh comets seen each year near the Sun, about one, and extrapolated the number on the basis of physical and dynamical simulations. He arrived at a population figure for the Oort Cloud of roughly 200 billion comets. More recent simulations have increased the proposed number of Oort Cloud comets, even by as much as an order of magnitude, to as many as 2 trillion. Not only do astronomers now know of a greater influx of long-period comets, but they also realize how effective Jupiter is in preventing the return of some of these bodies. The inner solar system, because of the gas giants, is poorer in long-period comets than we would otherwise be. Modeling of the mass of the Oort Cloud suggests an outer Oort Cloud (beyond 20,000 astronomical units) of 5 Earth masses and an inner Oort Cloud of about 5 times that amount.

What about the troublemakers? The objects that perturb Oort Cloud comets, sending them inward toward the Sun, have been sources of fascination ever since Oort himself considered them. Prior to 1970, no one knew about giant molecular clouds. Encounters with these massive structures must be quite rare. Estimates place them in the range of three or four every 100 million years. But over time, the effects add up. Over the history of the solar system, they're believed to be as significant as the effects of all passing stars.

The size and shape of the Oort Cloud are not precisely known, of course. But planetary scientists believe the cloud is an oblate spheroid — a sphere with the equatorial axis stretched wider than the poles — with the equatorial plane more or less parallel to the galactic plane. The Oort Cloud may extend as far away from the Sun as 1½ light-years, or about 40 percent of the way to the nearest star, the Alpha-Proxima Centauri system.

Astronomers believe that in the solar system's past, stars have sometimes passed through the Oort Cloud. They propose that when this happened, a star would in effect tunnel its way through the cloud. A star of the Sun's mass, moving at 20 km per second, would excavate a tube about 150 billion km wide (a distance 20 times greater than that between the Sun and Pluto), scattering the comets within that cylinder outward. In fact, astronomers believe that plunging stars, over the entire history of the solar system, have ejected some 10 percent of the comets that originally belonged to it.

Planetary scientists further point to the notion that such a star plunging through the Oort Cloud would cause comets to shower into the inner solar system. A star plunging through the inner edge of the Oort Cloud, where many comets presumably reside, could cause a huge infall of millions of comets into the inner solar system. These kinds of events may have never happened, but if they did, they probably took place in the early history of the solar system, when the Sun was still in the open star cluster in which it was born.

In an attempt to analyze the potential dangers of a star passing near or through the Oort Cloud, astronomers have produced computer models, based on the *Hipparcos* spacecraft data, attempting to pinpoint possible close encounters with stars in the future. One such study suggests that about 12 stars pass within 3 light-years of the Sun every million years. Assuming the Sun's current orbit about the center of the Milky Way and the galactic neighborhood around us stay the same, this would mean that something like 50,000 close encounters with stars have taken place to this degree in the history of the solar system. But perhaps some three-quarters of these encounters have occurred with red dwarf stars, which would have little effect on the cloud because they lack sufficient mass to stir things up as a solar-mass star would.

Other effects are even more significant in perturbing Oort Cloud comets than passing or plunging stars, however – and are stronger than passing giant molecular clouds. The gravity of the Milky Way's disk itself is the most significant influence on the cloud's member comets. The galactic tide, perturbations in the gravity of the Milky Way's disk, is significant, causing an inward step of the perihelia of comets, helping to push comets inward more effectively, and shoving them past the gravitational barriers of Jupiter and Saturn.

Another question confronts astronomers who contemplate the Oort Cloud. If so many comets within it are dispersed over time, then is it steadily losing its cometary population or is something helping to replenish the cloud? Astronomers think it's unlikely that comets could be captured from interstellar space, simply because of the energy required to change orbital dynamics. But they do believe the outer Oort Cloud could be restocked from a reservoir of comets existing in the inner Oort Cloud. It may be that passing stars push up the energy of these comets' orbits to move them outward.

When it comes to where comets live, however, the Oort Cloud is not the only game in town. A thousand times closer in lies another significant population of icy, small bodies of the solar system. In 1930 the long search for a suspected Trans-Neptunian Object (TNO) in the solar system finally came to an end when American astronomer Clyde W. Tombaugh (1906–1997), a midwestern farm boy who took up astronomy at Lowell Observatory in Flagstaff, Arizona, discovered the then-planet Pluto (it was reclassified as a dwarf planet by the International Astronomical Union in 2006).

After Pluto's discovery, it occurred to a number of astronomers that other, Pluto-like bodies may be strewn about the outer solar system. American astronomer Frederick C. Leonard (1896–1960) wrote that it could be likely that "in Pluto there has come to light the first of a series of ultra-Neptunian bodies, the remaining members of which still await discovery but which are destined eventually to be detected." Thirteen years later Irish astronomer Kenneth Edgeworth (1880–1972) proposed

Figure 6.4. Discovered by American astronomer Edward E. Barnard in 1889, Comet 177P/Barnard shows a large, diffuse coma during a recent apparition on August 30, 2006. The imager used a 20-inch Ritchey-Chretién telescope, CCD camera, and stacked exposures. Credit: Paulo Candy.

that the area of the solar system beyond Neptune offered too much space for the primordial solar nebula to condense planets, but instead produced numerous small solar system bodies. He suggested that occasionally some of these icy bodies could wander into the inner solar system as comets.

Most famously, in 1951 Dutch-American astronomer Gerard P. Kuiper (1905–1973) wrote a survey of the possibilities of this region in the journal *Astrophysics*. He suggested that a disk of icy bodies formed in this region in the early history of the solar system but that it probably didn't still exist today. Kuiper initially believed, like other planetary scientists of the day, that Pluto was as large as Earth and the mass in the disk was probably scattered outward.

Over the following years other tantalizing clues about the existence of distant, icy bodies occasionally cropped up. The 1977 discovery by American astronomer Charles Kowal (1940–2011) of the icy "asteroid" 2060 Chiron was one such clue. Chiron was unlike any object previously known, with an orbit that carried it out into the region between Saturn and Uranus. It became the most distant asteroid known. It was given the designation of an asteroid, but later, in 1988, Chiron seemed to behave as a comet, showing sudden brightening and a coma. Thus, Chiron also received the cometary designation 95P/Chiron. The previously clean line between comets and asteroids was suddenly becoming quite blurry.

Astronomers subsequently found that Chiron spans about 230 km, rotates about every 6 hours, and has a spectrum similar to that of the nucleus of Halley's Comet. It turns out that Chiron was the first of a class of objects with orbits showing a perihelion beyond the orbit of Jupiter and a semimajor axis smaller than that of Neptune. Thus, these bodies live among the giant planets of the solar system. Because they demonstrate characteristics of both asteroids and comets, these objects are called

Centaurs – after the half-horse, half-human figures in Greek mythology. Astronomers believe that perhaps as many as 44,000 Centaurs with diameters larger than a kilometer exist.

The first Centaur to be discovered – although it wasn't known to be a Centaur until decades later – was 944 Hidalgo, discovered in 1920. In 1992 astronomers discovered another object similar to Chiron, 5145 Pholus, a 185-km body that has not shown cometary activity. The largest Centaur is 10199 Chariklo, measuring 260 km across, about the size of a main-belt asteroid. Three Centaurs have shown clear indications of cometary activity, Chiron, 60558 Echeclus, and 166P/NEAT. It's likely that Saturn's moon Phoebe is a captured Centaur. At present, 20 Centaurs are named.

From the initial analysis of Chiron's orbit, it seemed clear to planetary scientists that the orbits of these Centaurs were pretty unstable. And if that is the case, then something must be reinforcing this population of icy objects. Astronomers also found that short-period comets in the inner solar system were too plentiful to be supplied simply from the Oort Cloud. Moreover, short-period comets tend to cluster near the ecliptic plane of the solar system, while Oort Cloud comets are distributed randomly. Further hypothesizing about a disk of icy bodies beyond Neptune, Canadian astronomer Scott Tremaine (1950–) penned a paper with colleagues in 1988 in which he used the phrase "Kuiper Belt" to describe this disk. The phrase stuck, although sometimes the disk is called the Edgeworth-Kuiper Belt.

But the discovery of a verifiable Kuiper Belt object had not yet occurred. During the same period as the publication of Tremaine's paper, English-American astronomer David Jewett (1958–) focused his research on finding other bodies beyond the orbit of Pluto. Enlisting one of his graduate students, Vietnamese-American astronomer Jane Luu (1963–), to help him, Jewett commenced a systematic search for distant Pluto-like objects with telescopes at Kitt Peak National Observatory in Arizona and the Cerro Tololo Inter-American Observatory in Chile. They soon moved to the Institute for Astronomy at the University of Hawaii. After searching for 5 years, Jewett and Luu found a strange object on August 30, 1992, and subsequently announced the discovery of the first verified member of the Kuiper Belt, (15760) 1992QB$_1$. It is a roughly 160-km-diameter object that orbits some 41 to 46 astronomical units from the Sun, in the same region as Pluto.

Six months later, Jewett and Luu discovered the second Kuiper Belt object, (181708) 1993 FW. The hypothesized Kuiper Belt was coming to life. Astronomers stepped up their searches for other objects in the region and the names "Kuiper Belt" and "Edgeworth-Kuiper Belt" were born, although Brian Marsden made clear that neither Kuiper nor Edgeworth really envisioned a region that was quite like what astronomers were finding, and that Fred Whipple had. Jewett looked back to the 1980 prediction of a comet belt in the region by Uruguayan astronomer Julio Fernandez (1946–) as the most accurate one.

In any case, objects in the Kuiper Belt soon became known in greater numbers and known as KBOs – Kuiper Belt Objects. Clyde Tombaugh suggested the name "kuiperoids," which has sometimes been used. After some time passed, however, the now-accepted term Trans-Neptunian Object arose and has been widely used in all recent works. This encompasses all solar system objects that orbit the Sun at a greater distance than Neptune, so it includes asteroids and comets as well as Pluto (the first TNO to be discovered); dwarf planets Eris, Makemake, and Haumea; and the weird asteroidal object Sedna.

More than 80 years after the discovery of Pluto and 20-plus years after Jewett and Luu found 1992QB$_1$, astronomers now know of more than 1,200 TNOs. Although astronomers are in the relative infancy of discovering objects in the region, some interesting conclusions have resulted from various research projects. Early on, multiple studies demonstrated that the Kuiper Belt itself is not the source of short-period comets, but a linked population called the scattered disk is. The scattered disk is an unstable population of distant icy small bodies that have been scattered by the gas giant planets. They can approach to within about 30 astronomical units of the Sun, but also extend out to some 100 astronomical units. The innermost region of the scattered disk overlaps the Kuiper Belt. Planetary scientists believe the scattered disk formed when Neptune migrated outward into the still-forming Kuiper Belt, which was then closer to the Sun, creating the stable Kuiper Belt and the unstable scattered disk, whose objects can be perturbed by Neptune.

But the definitions of behavior in these regions and of membership are not clean and simple. Over what time scales do planetary scientists observe TNOs to include them either in the Kuiper Belt or as members of the scattered disk? Such icy objects can orbit, trapped for periods of time in resonances, before they change their dynamics and move in and out of the scattering population – perhaps many times over millions or billions of years.

So planetary scientists really would like to link the scattered disk with the mechanism by which it formed. A more precise definition of the scattered disk, then, might be to suggest it is the region of space inhabited by TNOs that have encountered Neptune within a Hill's radius – that is, within about 116 million km of the planet, enough to be gravitationally affected by it – at least one time during solar system history. The Kuiper Belt, then, is the region of space that complements the scattered disk and lies at a distance of at least 30 astronomical units from the Sun.

The question then arises, How did the objects within both the Kuiper Belt and the scattered disk form in the first place? Probably all bodies within the proto solar system formed in orbits with small eccentricities and with small orbital inclinations, as suggested by computer models of accretion. TNOs now in the scattered disk population may have begun with orbits that were more or less circular and placed them in the region of Neptune. While scattered disk objects don't provide astronomers with any clues about the early architecture of the solar system, Kuiper Belt Objects might. Their orbits suggest they were "powered up" into high eccentricities and inclinations early on and so may reveal something about a process that was at work in the early solar system but no longer occurs.

Astronomers are particularly intrigued by distant Kuiper Belt Objects on highly eccentric orbits that are farther away than 50 astronomical units from the Sun. Such highly energized orbits suggest some real dynamical fun happened in the past. They cite 2000 CR_{105} as a particularly intriguing TNO. It moves in a highly inclined (22.7°) orbit, oscillating between 44 and 230 astronomical units from the Sun. How could an object have been "spun up" into such a strange orbit as this?

An object like 2000 CR_{105} is termed an extended scattered disk object because it's actually outside the parameters of the scattered disk. Planetary scientists also believe an object like this probably formed much closer to the Sun and migrated far outward into its present orbit. And it's possible that a large number of extended scattered disk TNOs have yet to be discovered, and the known population will grow dramatically as more and more objects are found.

The scattered disk contains two discrete populations, and the Kuiper Belt actually has more than one orbital class of TNOs too. Among KBOs, objects that show orbital resonance with Neptune – that is, Neptune and the KBO have a regular, periodic, gravitational influence on each other – are called resonant TNOs. The dominant resonant population is the 2:3 resonance class, at 39.4 astronomical units from the Sun, with periods of about 250 years. These objects are sometimes called plutinos and consist of 92 members, although another 104 are suspected but unconfirmed. For every two orbits of the Sun a plutino makes, Neptune makes three orbits. Pluto itself is a plutino (thus the name), as well as 90482 Orcus, 28978 Ixion, and 38628 Huya.

A population of 10 resonant TNOs exists at an orbital distance of 42.3 astronomical units and with a resonance of 3:5. These objects orbit the Sun approximately every 275 years. This class includes (126154) 2001 YH_{140}, (15809) 1994 JS, and (143751) 2003 US_{292}. Another class lies at the distance of 43.7 astronomical units and has a resonance of 4:7, orbiting the Sun roughly once every 290 years. The 20 members of this resonance include 1999 CD_{158}, (119070) 2001 KP_{77}, and (118698) 2000 OY_{51}. The next class, "twotinos," have resonances of 1:2 and orbit about once every 330 years at a distance of 47.8 astronomical units. This class represents the so-called outer edge of the Kuiper Belt and contains 14 objects, including (119979) 2002 WC_{19}, (137295) 1999 RB_{216}, and (20161) 1996 TR_{66}. A 2:5 resonant class exists at 55.4 astronomical units, with an orbit of about 410 years. This includes (84522) 2002 TC_{302}, (119608) 2001 KC_{77}, and (69988) 1998 WA_{31}.

The many subcategories of resonant KBOs aside, Kuiper Belt residents without resonances are called classical KBOs, or cubewanos (queue-bee-one-Os), after the first such object found, 1992 QB_1. Objects with a Neptunian resonance are "protected" by Neptune in terms of stability, and classical KBOs are not. So that means the ones that stick around – by not encountering Neptune too closely – have moderate orbital eccentricities and usually are more distant from the Sun than 35 astronomical units. But there are a few exceptions to the rule.

When Gerard Kuiper envisioned his region of small bodies in 1951, he wondered why so much mass would be ordered in the solar system from Jupiter through Saturn, Uranus, and Neptune, and then essentially drop off to practically nothing. But thus far observers have found a relatively small amount of mass in the Kuiper Belt – about two orders of magnitude less than Kuiper himself had expected.

Astronomers still believe that long ago a primordial Kuiper Belt held far more mass than it apparently does today. In 1995 American astronomer Alan Stern (1957–) found that objects now in the Kuiper Belt could not have formed there, as collisions that would build up objects could not have taken place frequently enough over the age of the solar system. He also found that because of high orbital inclinations and eccentricities and fast impact velocities, collisions in the Kuiper Belt fragment objects rather than accreting them. They make small pieces out of larger ones, rather than the reverse.

Computer modeling reinforced the idea that the early Kuiper Belt was a massive storehouse of icy bodies. So where did so many of them go? From the perspective of a detective story, at least part of the answer is a little disappointing. Stern and other astronomers ran analyses of this question in the late 1990s and found that collisions in the Kuiper Belt may have ground down KBOs to essentially dust particles, which can then be carried off into the more distant solar system by radiation pressure or by the Poynting-Robertson effect, which causes dust particles ultimately to be dragged into the Sun. At modest inclinations and eccentricities, the collisions would have reduced the mass by a modest amount. But at higher inclinations and eccentricities, like the ones commonly observed now, many of the KBOs probably were destroyed outright, pulverized into dust that was then driven away.

Planetary scientists suspect that 99 percent of the original mass of the Kuiper Belt is no longer there. The collisional grinding scenario is a strong one for understanding how so much mass could have been lost, but not everyone is sold on it. For example, the fact that Pluto and its massive moon Charon form a double object suggests the collision of two similar-sized bodies that would be the largest ones known in the Kuiper Belt. How likely is it that two such bodies would have collided? And the existence of binary KBOs with wide separations between the two gravitationally bound objects also argues against massive numbers of grinding collisions.

Another unusual set of clues about the early history of the solar system comes from the so-called excitement of Kuiper Belt Objects. Planetary scientists expect objects in the scattered disk and those with resonances have orbits that were affected by Neptune. But the classical KBOs should not have high inclinations and eccentricities, and many do – they were somehow "excited" into higher-energy orbits, a strange outcome given what astronomers know about the formation of the solar system's primordial disk.

KBOs show different types of unusual orbits. Objects with high eccentricities have orbits that carry them closer to the Sun, making them brighter and easier to observe at perihelion. Objects with high orbital inclinations spend relatively little time at the latitudes where many surveys take place, making them a little harder to detect. Conversely, objects with low inclinations spend small amounts of time at high latitudes, where other searches have taken place.

Analyses suggest that among the classical KBOs, two different populations exist – those at high inclinations and those at low inclinations. Astronomers call classical KBOs with inclinations greater than 4° "hot" objects and those with inclinations less than 4° "cold" KBOs. The dynamics of separating these objects into the two populations are still being debated and discussed. Either a number of the objects that initially had low inclinations were excited into high-energy orbits, or two populations formed in a distinct way and the hot population has always existed in such orbits.

A tantalizing clue about the dynamics of the hot KBOs is found by studying their physical nature and how they may differ from cold KBOs. In 2001 several astronomers noticed that intrinsically brighter classical KBOs tend to be found at high orbital inclinations. Also, high-inclination classical KBOs tend to be less red than their low-inclination counterparts, a characteristic that somehow relates to their chemical surface compositions. The hot classical KBOs resemble other types of Kuiper Belt Objects in terms of color, while the low-inclination classical KBOs appear to be different.

Planetary scientists still have much to learn about the Kuiper Belt, and about the nature of objects within it. But it's clear that this inner major storehouse of icy bodies holds tens of thousands of objects, most of which await discovery. It's known that the Kuiper Belt has an edge at a distance of 50 astronomical units from the Sun, and astronomers believe this edge may have been created by an early encounter with a star that passed some 150 to 200 astronomical units away from the Sun. Moreover, they believe the existence of the odd orbit of 2000 CR_{105} could be explained by a stellar encounter at a distance of some 800 astronomical units from the Sun.

As we've seen, however, comets in the solar system are not confined to the Oort Cloud, the Centaur population, the scattered disk, and the Kuiper Belt. Short-period comets, those with periods of less than 200 years, live much closer in toward the Sun. They generally orbit close to the plane of the ecliptic, in the same direction as the planets, and travel out to aphelion distances of Jupiter and slightly beyond. Short-period comets subdivide into the two primary groups, the first of which are the Jupiter-family comets – those with periods shorter than 20 years and inclinations of 20° to 30°. Currently around 500 Jupiter-family comets are known; they include all

Figure 6.5. In late 2012 the normally faint Comet 168P/Hergenrother underwent a sudden explosive brightening, gaining six magnitudes within a brief time, and displaying a luminous coma and faint, broad tail. On October 8, 2012, the imager used a 12.5-inch f/5 reflector and a 60-minute exposure at ISO 800. Credit: Chris Schur.

Figure 6.6. Comet Cardinal (C/2008 T2) passes the large open star cluster M38 in Auriga (bottom) and more compact open cluster NGC 1907 (top) on April 14, 2009. The imager used a 180 mm astrograph, a CCD camera, and stacked exposures. Credit: Brian Kimball.

manner of periodic comets very familiar to backyard observers: 2P/Encke, 6P/d'Arrest, 9P/Tempel 1, 46P/Wirtanen, and 67P/Churyumov-Gerasimenko among them.

The other major class of short-period comets is the Halley-type group, comets with periods of 20 to 200 years and with inclinations of 0° to more than 90°. This family includes around 70 members, including the namesake 1P/Halley along with other familiar names: 23P/Brorsen-Metcalf, 55P/Tempel-Tuttle, 109P/Swift-Tuttle, and 161P/Hartley-IRAS.

The unusual class of comets known as Main-belt comets arose at the end of the 1970s with the discovery of an odd minor planet, 7968 Elst-Pizarro. Discovered by

Belgian astronomer Eric Elst and Chilean astronomer Guido Pizarro, the object at first appeared to be a garden-variety minor planet but soon showed signs of cometary activity. The object now has the designation 133P/Elst-Pizarro. Main-belt comets like this one have nearly circular orbits within the main asteroid belt that are nearly indistinguishable from those of asteroids. These objects differ from short-period comets by having small inclinations and low eccentricities. Eight Main-belt comets are known: Elst-Pizarro, 176P/LINEAR, 238P/Read, 259P/Garradd, P/2006 VW139, P/2010 R2 La Sagra, P/2012 F5 Gibbs, and P/2012 T1 PANSTARRS.

And of course some comets are among the near-Earth objects (NEOs) that pass close enough to Earth's orbit to pose a potential threat. Of the approximately 10,000 NEOs known to planetary scientists, 93 are comets and the rest asteroids. Any of these 93 comets could indeed intersect our planet's orbit in the future, recreating the horror of what took place in Siberia in 1908 – probably a cometary airburst – or the result of uncountable collisions of comets into Earth in the early days of the solar system. Which raises an interesting question: Just what exactly did comets carry along with them to our young planet?

7

The Expanding Science of Comets

The science of understanding comets has changed dramatically in the 60 years since Fred Whipple hypothesized the basic model of cometary nuclei and Jan Oort suggested the existence of the large cloud that most comets call home. Over the past 20 years some really interesting twists and turns in the road have confronted planetary scientists: The distinction between comets and asteroids has blurred significantly and is based in some ways on where these icy bodies happen to be at the present time rather than intrinsic differences in their makeup.

The past decade has also witnessed an explosion of science, conjecture, and speculation about the role comets have played in our own existence. Did comets provide substantial amounts of water to Earth, thereby making the development of life on our planet either possible or easier? And even more dramatically, astronomers have focused on the organic molecules discovered in comets for years now, wondering whether they may have deposited organics onto early Earth that may have led to the development of life. We've seen that comets and asteroids are destroyers of life when they cause large impacts on our planet. Were they also the promoters of life? Do we owe our very existence to them?

For some years beginning in the 1990s it became trendy for scientists and laypeople to jump on the bandwagon of the idea that comets supplied huge amounts of water to early Earth. Conferences included considerable talk about the notion. Various writers pounded out books suggesting the idea was ironclad. Several studies cropped up adding apparent weight to the evidence that much or most of Earth's water rained down from outside the atmosphere when Earth was in its formative years.

At present, however, it seems to be one of those case studies where lots of people draw sweeping conclusions based on a few studies, a few meetings, a few books. That's a disease that permeates culture these days – believing that one study warrants

Figure 7.1. On September 30, 2010, Comet 103P/Hartley 2 made its way past the emission nebula NGC 281, creating a stunning pairing for astroimagers. This view was made with a 106 mm refractor at f/3.6, a CCD camera, and stacked exposures. Credit: Don Taylor.

a concrete, permanent conclusion. But that's not the way science works: Instead, it is a slowly developing, enormous machine with numerous researchers and studies grinding away at the best conclusions. They develop over ample time frames as data are gradually refined. The last decade has pushed the idea that lots of Earth water originated from comets into a relatively weak position.

To investigate the possibilities, planetary scientists want to know as much as they can about the conditions on early Earth. We know that Earth was cooler in its early history, and that the young Sun produced about 40 percent less radiation than it does today. With such a faint early Sun and a cool early Earth, what were conditions like on our planet?

It's difficult to gain a window billions of years back into the planet's history. But scientists can draw on some very stable mineral specimens with evidence of conditions on early Earth trapped inside them.

Zircon, zirconium silicate, $Zr(SiO_4)$, is an extremely stable mineral over long intervals of time that also contains tiny amounts of radioactive impurities like uranium that enable dating the crystallization ages of individual grains (by radioactive half-lives). Zircon samples from the earliest days of Earth allow scientists to examine isotope ratios of oxygen trapped within the crystals to understand early Earth conditions. They find that surface temperatures were low and that oceans would have existed rather than a thick, steam-rich atmosphere. Zircon samples analyzed from Jack Hills, Western Australia, for example, from about 4.4 billion years ago, suggest interaction with low-temperature liquid water in a surface environment. (Jack Hills represents the oldest continental fragments on Earth.) This picture is radically different from the ages-old idea that early Earth was an extremely hot place covered with liquid magma.

So if early Earth had lots of liquid water, in oceans no less, then where did the water originate? Certainly the earliest period of Earth as a planet saw innumerable impacts of comets and asteroids into our planet. The inner solar system was far more densely populated with small bodies in the early years, and these bodies were "cleared out" mainly by running into other bodies, as well as eventual gravitational perturbations. Planetary scientists hypothesize an important period in this early, dangerous, violent period of the inner solar system as the Late-Heavy Bombardment, which took place some 4.1–3.8 billion years ago. (The proposed idea that a Mars-sized body, Theia, struck Earth and led to the formation of the Moon suggests that happened well before the Late-Heavy Bombardment, some 4.5 billion years ago.)

The Late-Heavy Bombardment is called "late" because it occurred after most of the early impacts of the formative inner solar system. Support for the bombardment is found in the *Apollo* lunar samples, which show that most impact melt rocks on the Moon formed in this rather narrow window of time. Researchers in the 1970s noticed this clustering of ages and hypothesized an intense period of bombardment on the Moon (and of course elsewhere nearby) during that period.

The idea has gained ground as more lunar meteorites have been recovered. The approximately four-dozen meteorite falls traced back to a lunar origin presumably were from random sites around the Moon, unlike the *Apollo* samples, which were collected from three specific impact basin areas. Studies of the Moon and of Mercury suggest the same family of projectiles struck both bodies during the Late-Heavy Bombardment.

Of course, unlike Mercury and the Moon, Earth covers up its scars by a variety of erosional and tectonic processes. But extrapolating the numbers of lunar craters suggests that during the Late-Heavy Bombardment our planet was hit by a flurry of meteoroids that created about 22,000 or more craters of 20 km or larger, perhaps 40 impact basins with diameters around 1,000 km, and several impact basins with diameters of 5,000 km. No one knows whether any life existed on Earth at the time; the earliest microbial fossils date to about 3.4 billion years ago. If it did, those days would have been unimaginably violent ones to live through.

Because comets contain large amounts of water and other ices, the notion that bombardments by comets deposited much of Earth's water has seemed to be accepted almost as a measure of faith. But a variety of recent studies view this idea in a somewhat harsh light. And understanding how Earth got its water is certainly one of the most important unresolved questions about the early solar system.

In tracing where the water originated, scientists attempt to recreate the conditions of the protosolar nebula. They agree that the nebular disk was hotter and denser toward the center and cooler and less dense away from the center. The varying degrees of temperature throughout the protosolar disk clearly affected where water and icy particles existed. The central region would have contained high concentrations of metals and silicates, whereas icy particles could have existed in far greater quantities away from the center. They also believe the earliest solid particles were tiny; these objects accreted into larger ones by sticking together through countless collisions. Where plentiful oxygen existed, carbonaceous chondrite meteorites, which can contain up to 10 percent water, formed. But comets, on the icy perimeter, contain as much as 80 percent water by mass.

As hard as it is to believe when one stands on the shore of a great ocean, Earth has a small amount of water by mass - only 0.02 percent in its oceans and a little more than that below ground on continents. Despite the small fraction of water on Earth compared to its total mass, our planet has plenty of water. For a planet at our distance from the Sun, it is exceedingly rich in water, containing far more than might exist here.

So how did the water get here? The possibilities include bombardment by comets, asteroids, and planetesimals; water absorption by silicate grains in the protosolar nebula and transport into early Earth; and the production of liquid water through oxidation of a hydrogen-rich atmosphere. A significant recent clue to the likelihood of each of these scenarios comes from studying the ratios of deuterium to hydrogen from each of them as well as predictions made by computer modeling of how much water each of these methods would produce.

A years-long affair with the idea of comets delivering a huge amount of water to Earth seemed built of pure, simple logic. Made largely of water ice, and existing perhaps in the trillions, they were the leading suspects. They also presumably retained their isotopic properties from the earliest days of the solar system. But recent measurements of the deuterium to hydrogen ratio (D/H ratio) of water in eight Oort Cloud comets deliver a heavy blow to this idea. Water can carry different isotopic signatures, depending on the hydrogen isotopes it contains, and the ratios of deuterium (heavy hydrogen) to hydrogen in the comets are on average twice those of the Vienna Standard Mean Ocean Water (VSMOW), which defines the average isotopic water concentration on Earth. Moreover, they are 15 times greater than the D/H ratio of the early solar nebula. Although it's quite possible the D/H ratio of ocean water could have changed over time, most scientists believe this incompatibility rules out comets as major sources of Earth's water.

But the arguments over comets providing water to early Earth didn't stop with the Oort Cloud. In the 1990s, several planetary scientists proposed that perhaps Jupiter-family comets were the source of Earth's water. These comets have lower D/H ratios because they formed in a warmer region of the solar system. For example, the Jupiter-family Comet 103P/Hartley 2 has nearly the same D/H ratio as Earth's ocean water. But dynamical arguments stepped in to make the Jupiter-family comets unlikely as sources of water. With Jupiter and Saturn in their current positions, comets

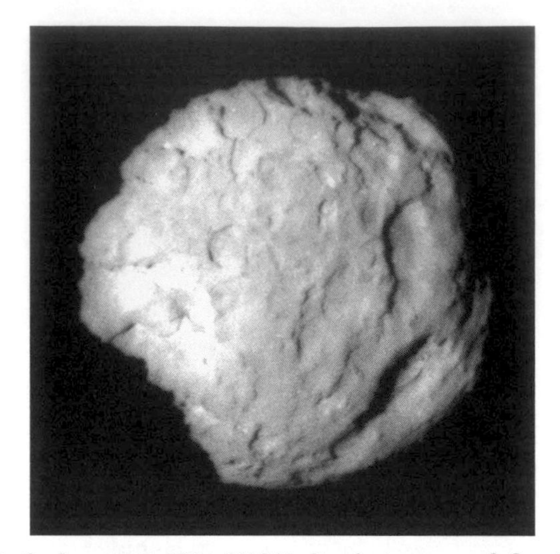

Figure 7.2. On January 2, 2004, NASA's Stardust spacecraft flew just 237 km from the surface of Comet 81P/Wild 2, revealing this heavily pockmarked surface of the nucleus, which spans some 5.5 by 4.0 by 3.3 km. Credit: NASA/JPL-Caltech.

that bombarded Earth may have traveled from the region beyond Uranus. But computer modeling of the early solar system shows a "bombardment" of comets from this region likely would have produced only about 6 percent of the water needed to make up Earth's oceans.

A spate of recent studies support this idea that comets played a relatively minor role in delivering water to early Earth. Several analyses show that comets probably contributed about 10 percent of Earth's water. But if the dynamics of orbits changed drastically over time, then this answer may be misleading. Support for the hypothesis that lots of water was supplied by comets in studies of isotopic ratios is also weak. Studies of ratios of noble gases suggest comets could not have delivered the majority of water to early Earth unless it happened very quickly, during something like the first 100 million years of the planet's history – or perhaps if comets came from a region other than the Oort Cloud. These studies, coupled with dynamical simulations, suggest that comets contributed 10 to 15 percent of the water on Earth.

Could primitive asteroids have delivered a significant additional amount of water? Carbonaceous chondrites do have D/H ratios quite similar to Earth's ocean water. That match might suggest that these primitive asteroids, which could have been more hydrated in the past than now, might be a mechanism for delivering water. And relatively water-rich asteroids from far out could have been perturbed inward toward Earth by the giant planets. But the dynamics of orbits, once again, gets in the way. The efficiency of scattering asteroids toward Earth is very low and that the contribution in water to our planet must have been very small. If primitive asteroids had 10 percent water by mass, the likely rate of accretion would have required the mass of asteroids to be 4 times that of Earth in a region some 2.5 astronomical units away – a figure that seems unrealistically high.

The timing of the delivery of water is also a problem. The scenario that primitive asteroids delivered lots of water suggests that it would have arrived on Earth very early on – when the planet was young, still accreting, and only 60 percent of its current mass. This would have been before numerous major impacts would take place, with planetesimals and perhaps several small planet-sized bodies. These traumatic impacts would have made it very difficult for the planet to retain this water if it was deposited by carbonaceous chondrites.

But there is a leading idea about the mechanisms to deliver significant water to early Earth. It seems clear that water must have been deposited on Earth during a long and sustained period during the planet's history, not just once, early on. Planetary scientists believe that the widespread impacting of water-bearing planetesimals and protoplanets into Earth probably delivered substantial amounts of water to our planet. These objects probably originated in the outer asteroid belt, which therefore becomes the leading suspect as the source of water in the solar system.

But delivery by water-rich asteroids is not the only method that perhaps played a significant role. Large amounts of water could also have come directly from the solar nebula. Water molecules can adhere to dust grains so efficiently, even at relatively high temperatures, that significant amounts of water – perhaps even the entirety of the oceans and more – could have existed in and on dust grains that accreted in massive numbers to form Earth. And the cooler the temperatures, the greater the efficiency of the water molecule adhesion to dust grains.

But there are questions and problems here too. The D/H ratio of the solar nebula does not match the current ocean water well. And planetary scientists can't currently explain easily how the water would be retained as dust grains accreted into larger and larger particle sizes.

So the issue of where Earth's water originated is unresolved, but it seems that a large percentage of it may well have arrived from water-bearing asteroids from the outer asteroid belt, supplemented by comet bombardment and by water molecules stuck to dust grains from the solar nebula. And the research goes on.

Well before the current rounds of investigating the origin of earthly water, scientists began to speculate about comets and organic molecules, seduced by the possibility of a link between comets and life.

The idea of such a link between comets and life really gained momentum in the 1970s, when an amazing and somewhat controversial figure in astronomy took up the issue. English astronomer Sir Fred Hoyle (1915–2001) was one of the more colorful figures of his time. He inadvertently invented the term "Big Bang" during a radio show in which he espoused his strong opposition to the idea, instead promoting the steady-state theory of the universe in which all matter falls back onto itself and undergoes recurrent big bangs. His belief in the steady-state theory would only show itself to be wrong during his later years. Before that, in the 1970s, Hoyle had another revolutionary idea.

Along with one of his graduate students, Sri Lankan astronomer Chandra Wickramasinghe (1939–), in 1974 Hoyle was the first to propose extraterrestrial abiogenesis – that is, the notion that life originated in space and was deposited onto Earth, rather than originating on the planet (or more specifically, in its oceans). Confronting the scientific mainstream (much as he had in opposition to the Big Bang), Hoyle proposed that life began in space itself, spreading through panspermia, and is distributed here, there, and everywhere by means of comets, asteroids, meteoroids, and cosmic dust.

The notion was not exactly original – the idea had been discussed as early as the 5th century B.C. by the Greek philosopher Anaxagoras (*ca.* 510–428 B.C.). Centuries later, notable scientists also considered the idea, including Swedish chemist Jöns Jacob Berzelius (1779–1848), British physicist William Thomson, Lord Kelvin (1824–1907), and German physician and physicist Hermann von Helmholtz (1821–1894).

The first to popularize the idea widely, Hoyle and Wickramasinghe noted that interstellar dust is carbonaceous, that is, organic, and was the seed for primitive living microbes (and even viruses) that continuously enter Earth's atmosphere, as well as permeate space through individual galaxies. They suggested that interstellar dust could have delivered life to Earth to begin with, and that a constant flow of microbes could cause epidemics on Earth, a shifting buffet of diseases, and genetic mutations. In 2009 the distinguished English physicist Stephen Hawking (1942–) supported the idea.

Much like the clever ideas of string theory and multiverses, panspermia is hypothetically possible but at this stage completely lacks supporting evidence of any kind. And of course in the business of science, the burden of proof is on the hypothesis, not the other way around. Hoyle and Wickramasinghe suggested the methods of transport throughout the universe could be radiation pressure pushing small particles along or microbial life existing in rocks that are dispersed through accretion and dynamical evolution.

Could comets, specifically, have transported life to Earth or carried the raw materials of life onto our planet? Planetary scientists have known for many years that comets are rich in organic molecules, are coated with dark organics like polycyclic aromatic hydrocarbons (PAHs), and contain abundant CHON particles – those primitive particles composed of the light elements carbon, hydrogen, oxygen, and nitrogen. The laundry list of molecules observed in comets is long and includes numerous carbon-containing compounds. If panspermia didn't generate life throughout the cosmos, then perhaps microbial life formed extraterrestrially in rare cases and was

Figure 7.3. Comet LINEAR (C/2012 K5) appeared as a bright, stubby object of strong yellow color when this image was made on January 4, 2013, using an 8-inch f/2.8 astrograph, a CCD camera, and stacked exposures. Credit: Gerald Rhemann.

delivered here by comets. Or perhaps comets deposited the raw materials, the organics that led to higher-order organics, from which life arose.

And some respected scientists have even suggested that life arose in deep space and may have been sent spaceward, arriving on Earth or other planets on purpose, in a process that came to be called "directed panspermia." Those who either supported the idea or at least wrote about it in detail included American astronomer Carl Sagan and English molecular biologist Francis Crick (1916–2004), codiscoverer of the structure of the DNA molecule.

Before examining the evidence for the relationship between comets and life, however, it's probably a good idea to establish just what we mean by life. It's a trickier question than you might guess. And the history of ideas that attempt to define life is pretty fascinating in and of itself.

One of the great early definitions of life is from none other than our old Greek friend Aristotle, who proposed simply, "By life we mean self-nutrition and growth (with its correlative decay)." Nearly 2,400 years ago, the great philosopher had identified that organisms exchange energy with the environment and that life, overall, is a historical phenomenon.

Relatively little progress occurred with defining life for many centuries, but a burst of creativity struck various philosophers and writers during the late 18th and early 19th centuries. In 1802, French anatomist and physiologist Marie Bichat

(1771–1802) recorded, "Life is the totality of functions that resist death." That may sound a little self-evident, as echoed in the late comedian George Carlin's wisecrack "That's the whole secret to life … not dying!"

But the avoidance of death did underscore the search for other attributes of what makes something alive in the cosmos and led to considering the totality of living systems and everything they require to function. In addition to systems, French naturalist Georges-Louis Leclerc, comte de Buffon (1707–1788), and later English naturalist Charles Darwin (1809–1882) placed reproduction at the center of the definition of life. And Darwin's contribution of course led to the central engine of life as it relates to the universe over time, in the form of evolution. And in the 20th century, adding the increasingly sophisticated understanding of the chemistry of life, of DNA, of proteins, of amino acids, led to a greater incorporation of molecules in defining life.

Present-day science holds two sacred truths as key to understanding life. The first is that living beings are chemical systems able to maintain their existence by permanently synthesizing all or most of their components. The second is the ability to reproduce. These two factors are held as crowning principles in defining life by chemists, on one hand, and by geneticists and biologists, on the other.

With these two principles established, some characteristics of living things can also be specified. They constitute seven physiological functions that chemists and biologists agree help to define life in the cosmos: homeostasis, the regulation of an internal state to maintain balance; organization, as with cellular structure; metabolism, transformation of energy; growth, with creation of new matter outpacing the loss of old; adaptation, the ability to change in response to the environment; response to stimuli, reacting to events or occurrences; and reproduction, the ability to produce new individual organisms.

Understanding what defines life is only a first step toward exploring any role comets may have played in our existence. We also need to grasp the basics of how life probably got its start on early Earth, and a brief story of what happened to make us what we are today.

The best research today suggests that life originated on early Earth at least 3.7 billion years ago, just after the period of the Late-Heavy Bombardment. In 2011 a team of scientists published their analyses of what are currently the oldest life-forms known on Earth, microfossils of bacteria from the 3.4-billion-year-old Strelley Pool Formation in Western Australia. These primitive organisms are associated with tiny crystals of the mineral pyrite – iron sulfide – and metabolized sulfur. These microstructures exhibit hollow cell lumens, carbonaceous cell walls rich in nitrogen, and organization in chains and clusters. They are the oldest life known on our planet.

Exactly how life began on early Earth is not fully understood, of course, but scientists know the basic ingredients were all present in large quantities – from key elements like carbon, hydrogen, nitrogen, oxygen, phosphorus, and sulfur to abundant energy, warmth, electricity, magnetism, and attractive places for atoms and molecules to combine, such as plentiful liquid water oceans with abundant hydrothermal vents, called black smokers, that spewed forth a rich broth of hot organic molecules. Given that atoms combine in specific ways to form molecules and compounds, according to their electrical charges, life on early Earth seems to have been inevitable. It was simply a matter of where and when.

Models of how life originated build on several important milestones in thinking. One was the celebrated Miller-Urey experiment, conducted in 1952 at the University of Chicago by American chemists Stanley Miller (1930–2007) and Harold C. Urey (1893–1981). Simulating the conditions thought to exist on early Earth, the experiment sparked a chemical broth "soup" with a simulated lightning charge and synthesized more than 20 amino acids, the building blocks of proteins.

The key organic molecules, RNA (ribonucleic acid) and the more sophisticated molecule, DNA (deoxyribonucleic acid), contain the genetic codes of organisms. Along with proteins, these components are essential for life and work together to make it possible. Whether genes originated first and were followed by proteins, or vice versa, all of the chemistry for life was present in abundance in early Earth. (The two may have arisen simultaneously, as each needs the other.) Additionally, phospholipids formed from abundant elements and made up basic cellular membranes.

The uncertainty about whether genes or proteins originated first led some scientists (such as Francis Crick) to propose that original life-forms were based on RNA, which has some DNA-like properties as well as those of some proteins. DNA-based life may have evolved somewhat later on.

Once it was established, life appears to have taken a mighty foothold on the planet at a rapid pace. The two most critical of life's components, carbon and water, exist in great quantity and can be extracted from the environment, as with carbon dioxide. It's quite possible that the RNA world commenced life. Or perhaps it was originally an iron-sulfur world of organisms metabolizing iron sulfide just like those earliest microfossils discovered in Western Australia. Or perhaps life began as a lipid world, in which double-walled bubbles of lipids like those in cellular membranes arose in the oceans.

However it arose, life began as simple protobacteria and stayed that way for a very long time. (Viruses, discovered in 1892 and thought to be alive in the early 20th century, are really nonliving predators that attach themselves to living cells and overtake them, so are not independently living.) The first 1.5 billion years of life on Earth featured nothing more than simple prokaryotes, primitive microbes lacking a cellular nucleus. Some 2 billion years ago, the first multicellular organisms appeared. Many invertebrates came onto the scene. Some 1.5 billion years ago the first eukaryotes appeared – microbes with a cellular nucleus. And a buildup of oxygen in Earth's atmosphere made many other forms of life possible in the ensuing eons.

Within the last half-billion years, Earth's climate and its set of living organisms underwent dramatic change. The Cambrian period, beginning 570 million years ago, saw an explosion of the first large numbers of mineralized (and thus fossilized) large multicellular organisms such as trilobites and brachiopods. During the Ordovician period, beginning 520 million years ago, marine invertebrates including jawless fish prospered. The Silurian period, beginning 450 million years ago, introduced the first jawed fish and the first terrestrial arthropods. During the Devonian period, beginning 420 million years ago, fish diversified and early amphibians appeared. The Carboniferous period, beginning 375 million years ago, was the beginning of commenced reptiles. And the Permian period, beginning 285 million years ago, saw commonplace mammal-like reptiles and widespread extinction of amphibians.

Then a great transformation on Earth occurred during the Triassic period, which began 240 million years ago. The mammal-like reptiles met their end and were replaced by dinosaurs, creatures that by rights should have existed down to this day, if not for a certain earthbound asteroid. The Jurassic period, beginning 195 million years ago, introduced the first birds and great diversity among reptiles. Starting 135 million years ago, the Cretaceous period saw the extinction of ancient birds and reptiles.

Then the K-Pg Impact that killed off the dinosaurs and commenced the Tertiary period, 66 million years ago, occurred. The modern orders of mammals evolved further. Human beings evolved over the past 5 or so million years. Our closest ancestors evolved over the past 2 million years and led, starting roughly 100,000 years ago, to *Homo sapiens*. And here we are.

So what do comets have to do with the responsibility for the origin of life on Earth, let alone cats, dogs, sharks, bears, or people?

In 2009 astronomers took a significant step in answering this question. A team of NASA researchers announced the discovery of glycine, an amino acid, in samples recovered from the *Stardust* spacecraft from Comet 81P/Wild 2, which the spacecraft encountered in 2004. The announcement was exciting because glycine, as one of the amino acids, is a fundamental building block of life. The fact that it exists in a comet – indeed, the only comet astronomers have returned a sample of to Earth – is compelling evidence that amino acids are commonplace in the icy reservoirs of the outer solar system and were abundant in the solar system's early days.

Amino acids are critically important biological compounds made of groups of amines (NH_2) and carboxylic acid (COOH). The key elements of amino acids are carbon, hydrogen, oxygen, and nitrogen, the same elements existing in large quantities in comets as those primitive CHON particles. Some 500 amino acids are known; they are biologically critical because they are the basic components (monamers) that make up proteins. It's quite amazing to realize one of the key building blocks of life exists in a comet.

Figure 7.4. Comet Pojmański (C/2006 A1) shows a finely structured ion tail in this image made March 8, 2006, with a 400 mm lens at f/2.8, a CCD camera, and stacked exposures. Credit: Paulo Candy.

Glycine itself has the chemical formula NH_2CH_2COOH and is the simplest of the 20 amino acids commonly found in proteins. It exists as a colorless, sweet-tasting crystalline solid and was discovered in 1820 when French chemist Henri Braconnot (1780–1855) boiled gelatin along with sulfuric acid. Commercially, it's used as a sweetener and as a buffering agent in antacids, antiperspirants, cosmetics, and toiletries. It's also often added to pet foods and animal feed. Glycine is also used commercially in the production of rubber sponges, fertilizers, and metal complexants (substances that prevent precipitation of materials in solutions).

Glycine is not essential in human diets because the body biosynthesizes it using the amino acid serine, which is in turn synthesized from 3-phosphoglyceric acid. Once it exists within the body, glycine forms molecules used for a variety of functions, primarily to build proteins. It's also an important neurotransmitter in the spinal cord, brainstem, and retina. It's an important molecule for life.

The NASA team's discovery was exciting for those studying organic molecules in space because it supported the idea that some of life's components exist in space and were delivered by comets and asteroids to early Earth. Although amino acids like glycine could easily have also formed in the primordial soup on our planet, it would certainly have been convenient in terms of getting the stuff of life going to have them sent by special delivery.

On announcing their discovery in August 2009, scientists beamed with pride over the notion that it also helps support the idea that life is commonplace in the cosmos, as it certainly does. With the building blocks of proteins whizzing about solar systems here and there, slamming into planets, seeding life on many worlds is probable. All the worlds would need would be the right amounts of energy, the correct range of temperatures, and a solvent like water in order to start the magical process of making proteins, RNA, and DNA. And then the story of life begins anew, perhaps on countless worlds in the universe.

Consider, for a moment, the implications for life in the cosmos. If the building blocks of life are common everywhere, and we know chemistry and physics are uniform in the cosmos, then life could exist in staggering numbers of places. The numbers game makes for great cocktail party conversation, if nothing more. Our galaxy, the Milky Way, for example, contains something like 400 billion stars. (No one knows for sure because large numbers of stars are dwarfs that are hard to detect over large distances.) Astronomers know that about 125 billion galaxies must exist in the universe.

For a moment, we're obligated to mention the concept of cosmic inflation, the idea originated around 1980 by American astronomer Alan Guth (1947–) of the Massachusetts Institute of Technology. If inflation theory is correct, and most cosmologists believe it is, then the universe we see is by no means the entire universe. But for simplicity's sake, let's set aside cosmic inflation, even though we believe in it.

Simple multiplication tells us the universe contains a minimum of something like 50 thousand billion billion stars. That's an enormous number of potential places where planetary systems could host living beings. And our first steps looking out into the galaxy around us, really to just the nearest stars around us, have already uncovered more than 900 planets in more than 700 planetary systems orbiting other stars, although we can only really detect large planets quite close to us thus far. The numbers of planets in the universe must be almost uncountable. In January 2013, astronomers at the Harvard-Smithsonian Center for Astrophysics estimated that something like 17 billion Earth-sized exoplanets exist in our galaxy alone.

So the implications of comets spreading glycine and other biologically important molecules around are important. The 2009 announcement became the crowning achievement of data analysis from the *Stardust* mission. Proteins, the planetary scientists reminded us, are the workhorse molecules of life, utilized in everything from our hair to enzymes that govern our biological processes.

The capture of *Stardust*'s sample marks a high point in cometary research. Launched in 1999, the craft held a container something like a car's air filter made of aluminum that held a novel substance called aerogel, an incredibly light, spongelike material that is more than 99 percent empty space. As the spacecraft passed through the tail of 81P/Wild 2 on January 2, 2004, the aerogel captured tiny particles from the comet, and the grid containing it parachuted to Earth on January 15, 2006.

Studying the results occupied scientists for months. Incredible care had to be taken not to contaminate the samples, and the researchers were working with incredibly minute amounts of material. The scientists analyzed not only the aerogel and what it contained but also the aluminum sides of the chambers that held the aerogel in place. As molecules of gas from the comet passed through the aerogel, some of them stuck to it or to the aluminum.

Figure 7.5. Is this an asteroid or a comet? Discovered in 1997, Lagerkvist-Carsenty (P/1997 T3) was thought to be a minor planet until it began showing signs of cometary outgassing, faintly visible in this image made on October 6, 1997, with the 0.6-m Bochum Telescope at La Silla, Chile. Credit: ESO.

After they identified glycine molecules in the samples, the scientists took painstaking measures to rule out the possibility of contamination from the lab itself. The glycine could have been from people handling the samples or even during the manufacture of the spacecraft itself.

Planetary scientists ruled out contamination by studying the isotopes of carbon in the samples of glycine. The most common type of carbon atom, carbon-12, contains six protons and six neutrons. But carbon-13 is heavier, sporting an extra neutron. Scientists found larger amounts of carbon-13 atoms in the cometary glycine than in terrestrial samples; that finding made identifying the origin easier than it might have been.

Glycine wasn't the only surprise the *Stardust* mission uncovered in 81P/Wild 2. In late 2006 planetary scientists announced further findings that included an amazing mineralogical discovery. It shook the idea, slightly, that comets are pristine windows into the original state of the outer solar system. No doubt, 81P/Wild 2 spent the majority of its life in the Kuiper Belt, on the outer edge of the solar system beyond the orbit of Neptune. But in 1974 the comet encountered Jupiter, which perturbed it into its present, closer orbit.

Astronomers were amazed to find some material from the comet that clearly formed relatively close to the Sun. They discovered grains of a rare mineral, osbornite, titanium nitride, that forms at very high temperatures like 3,000 K. How could material like this, that presumably formed early in the history of the solar system very close to the Sun, be inside a comet that spent most of its life beyond Neptune?

The finding suggests that the early solar system was an incredibly violent and dynamically unstable place. Planetary scientists suggest the osbornite in 81P/Wild 2 implies that particles in the inner solar system could have been ejected violently in bipolar outflows that launched them perpendicularly to the ecliptic plane, shooting them high above the Sun's poles, with a subsequent rain falling back down onto the icy outer regions like the Kuiper Belt. It suggests the solar nebula was a far more violent place than astronomers had previously thought.

Analyses suggest the early solar system could transport particles as large as 20 microns from the inner solar system far outward toward the edge. (Twenty microns [micrometers] is about two-hundredths of a millimeter.) The osbornite suggests particles of this size could be launched on a wind that would carry them from a few solar radii and ejected at ballistic speeds outward, sending stuff that formed in a hot environment to the cold, icy depths. Most hot particles close in toward the Sun would have been silicate minerals like olivines and pyroxenes, magnesium iron silicates.

And *Stardust* found a variety of organic molecules that are richer in oxygen and nitrogen than the typical organics previously found in meteorites. They are giving scientists a good look at the pristine materials that survived intact from the formative days of the comet. The diversity of the organic compounds surprised scientists.

In 2010, scientists analyzing the *Stardust* material determined the first direct measurement of the age of cometary material. The results in this study were also surprising. Although comets are believed to be among the oldest objects in the solar system, researchers found by studying 81P/Wild 2 that inner solar system material was transported to the regions where comets formed at least 1.7 million years after the oldest solar system solids formed.

Among the *Stardust* grains studied for this part of the project were calcium-aluminum inclusions (CAIs), which are often found in primitive meteorites and are believed to be the oldest materials formed in the solar nebula. These objects formed in the inner region of the solar system, where temperatures were high.

That CAIs exist in Comet 81P/Wild 2 is another example of the tremendous, widespread mixing that was taking place in the early days of the solar system. This finding raised questions about the timing of the formation of comets relative to the inner solar system solids, and further research will be needed to sort out this complex topic.

All the analysis of cometary grains, from the rich organics like CHON particles, to the surprising minerals that formed in the inner solar system, to complex molecules like amino acids, leads to a way of looking at where comets are going. In a sense, comets that are active – that enter the inner solar system and warm up once in a great while – are like living beings. They are evolving, decaying bodies moving from a pristine state to a state of defunct existence, in which they are burned out bodies without volatiles or even outright dead – smashed and destroyed.

The behavior of comets, if you will, varies from group to group as well as from comet to comet. The orbital dynamics and fates of Oort Cloud comets are different from those of Halley-type comets, which in turn are different from the behaviors of Kuiper Belt, Centaurs, and Jupiter-family comets. A variety of factors influence how these comets live their lives, so to speak, and how they die.

The surface temperatures of comets in their normal environments are quite low, on the order of 10 K in the Oort Cloud and 40 K in the Kuiper Belt. As the orbit of a comet carries it toward the Sun, however, its temperature slowly rises and this induces the sublimation of volatiles on the comet's surface, producing a coma and/or tail. This is what defines a comet. As we've seen, some asteroids have similar orbits to some comets, and the only differentiation is the production of a coma or lack thereof.

The first ice to sublimate on the exposed surface of a comet is carbon monoxide (CO), which probably produces the comae observed around Centaurs as they are viewed in the middle of the solar system. But some Centaur comets display no coma because their surface volatiles are gone, presumably, or because their surfaces are coated with mantles of relatively dense, unreactive complex organics or silicates. (And these Centaurs are defined as comets principally because of their orbits.) Although carbon monoxide is the first ice to sublimate, water ice soon joins this process and creates a far more dramatic coma; that usually happens when a comet reaches about the distance of Jupiter from the Sun.

When a comet's nucleus begins to lose mass through sublimation, a number of effects take place. Its shape, its size, the way it rotates, and even its fate can be altered by losing gas and dust. Torques on the nucleus, the nongravitational forces astronomers struggled to interpret and predict for so long, can change the active areas of outgassing on the nucleus.

Astronomers' knowledge of surface mantles covering some portions of the nuclei of comets originated from observations of 1P/Halley and accelerated from ground-based observations in the 1990s. Astronomers believe some mantles are so-called rubble piles, mantles consisting of blocks of material too large to be lost by gas spewing from the comet. Even if they loosely bounce along the surface, they can inhibit the comet's sublimation. These mantles can be very thin and would form on short time scales. They would also be unstable and could be ejected if gas pressures were great enough.

Another kind of mantle, the irradiation mantle, may exist too. This contains material that is highly evolved – chemically transformed by prolonged exposure to radiation such as cosmic rays. Such mantles could form, planetary scientists believe, on time scales of 10 million to 1 billion years, shorter than the so-called storage times of comets in the Oort Cloud and the Kuiper Belt. Through this mechanism, the material on the outer shells of comets could be significantly altered down to a depth of a few meters or less. Irradiation breaks the normal chemical bonds that exist between molecules. This allows hydrogen to escape while the surface becomes richer in carbon and oxygen. Large amounts of carbon make the nuclei of comets dark.

Both kinds of mantles probably exist. Comets that never approach the Sun very closely probably retain their irradiation mantles. Near-Earth Objects probably have lost this mantle.

As comets encounter the inner solar system, they also lose mass. Smaller objects are depleted most quickly, of course. It's difficult for astronomers to make precise measurements of the sizes of cometary nuclei because when they're close enough for study, an active coma gets in the way. Also, the comets with large, active areas on their nuclei are brighter and therefore easier to discover and study than intrinsically fainter comets. While it seems clear that comets with diameters of a kilometer or more should lose mass at a rate that will keep them around for very long times, comets with diameters smaller than a kilometer will not. These small comets will lose their volatiles before their dynamical life is over, populating the solar system with a large number of dead comets.

In smaller comets yet, a wave of thermal conduction through the nucleus, as the comet heats up, can reach the core; that may explain why some small and intrinsically faint comets could burst with rapid rotation, fragmenting into pieces.

Astronomers need to produce rotational light curves of cometary nuclei to infer anything about their shapes. Most nuclei of comets are elongated in one axis compared with main-belt asteroids of about the same size. This probably results when the comets lose mass in an uneven way, a process that heats areas nonuniformly and produces a nonspherical shape.

The fact that mass loss in a comet's nucleus is uneven also means the spin period of the nucleus and its rotational state can change dramatically as the comet continues to orbit the Sun. In fact, the physics of the outgassing on a comet's surface suggests the force can produce rotational instability, which could shatter a comet outright.

Energetic rotation of a cometary nucleus could also produce friction within the material that could dissipate energy over long intervals. A collision in the Kuiper Belt that produces a 6-hour spin rate for a 2-km comet could keep that excited rotation going for some 10 million years, about the time required for the comet to move from the Kuiper Belt to the inner solar system. After that, the energetic spinning would begin to dampen away. Astronomers believe that most comets in the inner solar system are in rotationally excited states.

As comets move toward the inner solar system, how do active areas of outgassing evolve? They are linked to the state of the mantle, which prohibits activity. As a comet warms, however, an exposed area of ice on the surface could warm and sublimate, allowing a channel of warmth to form that deepens as the comet's motion continues. Such vents in the surface of a comet are the first areas to open "faults" in the mantle, allowing other areas to warm and to diminish the mantle.

The colors of comets are different in different types of comets, and they change as a comet warms or cools. Kuiper Belt Objects show a wide range of colors, from neutral to extremely red. Centaurs also show a wide spread of color, suggesting that their mantles are relatively minor. Once a comet moves inside the orbit of Jupiter, however, the story changes. There, the comet can begin to sublimate water ice, and these comets do not show the very red colors of outer comets. Objects that are likely dead comets show coloration virtually identical to that of active comets.

When a comet runs out of its reservoir of ice-rich volatile materials, it becomes what is termed a dead or dormant comet. Dormant comets lack the volatiles but may have ice-rich centers. Dead comets have no volatile materials whatsoever. Classifying these objects is still something of an imprecise science, as with some objects planetary scientists have precious few clues or observational history to draw on. Indeed, the observational properties of some dead comets literally overlap those of some types of asteroids.

Astronomers believe some comets may in effect evolve into asteroids. Some 93 of the NEOs are thought to be defunct comets. Dying comets with rubble pile mantles may degrade over time into objects indistinguishable from asteroids. The transition from comet to asteroid may include a prolonged period in which comets have only intermittent cracks in their mantles that release a very small amount of outgassing, revealing a brief, temporary return to cometdom.

Planetary scientists know of several comets with orbits that strongly resemble those of asteroids – 2P/Encke, 3200 Phaethon, 107P/Wilson-Harrington, and 133P/Elst-Pizarro among them. The famous periodic, Comet Encke, is one that was in the gravitational grip of Jupiter but broke free, now in a strange, asteroidlike orbit (which is the shortest-period orbit of any comet). Comet 107P/Wilson-Harrington showed brief cometary activity on its discovery in 1949 but has appeared like an asteroid ever since. Elst-Pizarro, originally thought to be an asteroid, has shown cometary outbursts since 1996. Phaethon appears to be associated with the Geminid meteor stream even though it has shown no cometary activity.

Moreover, a number of asteroids are known to have cometlike orbits and could be dead comets. They include 5335 Damocles, 15504 1998 RG_{33}, 20461 1999 LD_{31}, 3552 Don Quixote, 1997 SE_5, and 1982 YA.

In contrast to fading gently away, comets can die spectacular deaths. Some comets break up by tidal forces from the Sun, as with a number of Kreutz Sungrazers, or a planet like Jupiter, as with Shoemaker-Levy 9 (D/1993 F2) and 16P/Brooks 2, which broke up in 1889. A small percentage of the Centaurs will pass very close to Jupiter within their lifetimes. This could eventually send some "new" comets into the inner

Figure 7.6. The short period comet 45P/Honda-Mrkos-Pajdušáková brightened to delight observers when this image was made on September 29, 2011. The imager used an 8-inch f/30 astrograph, a CCD camera, and stacked exposures. Credit: Gerald Rhemann.

solar system. Most comets that break up, however, do so from nontidal forces and the reasons for this are unclear.

In addition to breakups, however, comets can completely disintegrate. Almost all the examples of this spectacular, comet-ending phenomenon exist from the observations of the *SOHO* spacecraft imaging sun grazing comets very close to the star. Planetary scientists believe that Comet LINEAR (C/1999 S4), which disintegrated in 2000, may have experienced a buildup of gas pressure from within, which set off the comet's nucleus like a bomb. This kind of explosive disintegration would be more common with small comets, like LINEAR (which measured perhaps 0.9 km across), than larger ones.

The final signals from disintegrated comets arrive in Earth's skies every night, although at some times of the year far more than others. They are meteor streams, the detritus from dead or dying comets that intersect Earth's orbit and fall into our atmosphere, the small particles ionizing and lighting up as "shooting stars." The majority of good meteor streams originate from long-period or Halley-type comets, which produce better meteor events than those from Jupiter-family comets or those with orbits similar to those of asteroids.

Some of the better showers are those from familiar objects. The Quadrantid meteors, active in January, originate from the asteroid – dead comet – 2003 EH_1. The celebrated Perseid shower in August derives from Comet 109P/Swift-Tuttle. The Orionids, active in October, originate from the most famous of all comets, 1P/Halley.

The Geminids of December are linked to the orbit of asteroid 3200 Phaethon, another suspected dead comet. The Leonids, so attractive in November, are from Comet 55P/Tempel-Tuttle.

No one yet knows exactly how long these streams of debris from once-mighty comets will last. A study of the orbital dynamics of the Perseid stream suggests a lifetime of something like 40,000 to 80,000 years. For the time being, at least, when no bright comet appears in Earth's skies, we can at least go outside, look up into dark heavens, and find the small particles these comets once shed, raining down into our cosmic home on planet Earth.

8

Observing Comets

One of history's great lovers of comets is now remembered for just about anything but comets. French astronomer Charles Messier (1730–1817), who was born in Badonviller, Lorraine, and died in Paris, spent much of his career observing comets at a critical time in the history of observational astronomy. Messier was lured into astronomy by the excitement of seeing the Great Comet of 1744, with its multiple tails, and by observations of an annular eclipse of the Sun in 1748. And it was the return of Halley's Comet in 1759 that played a critical role in pushing Messier forward into his studies and cataloging of comets and cometlike objects.

In 1750, only about 50 comets were well known since the beginning of time. Keeping track of comets and studying their motions were the critical assignments for anyone interested in the subject. Messier traveled from his hometown to Paris in 1751, seeking his fortune, and was hired by astronomer Joseph-Nicolas Delisle (1688–1768) as a draftsman and recorder of observations. Although Messier first set about copying a map of the Great Wall of China, he was also instructed in the use of astronomical instruments. Messier also took a position as clerk at the Marine Observatory in Paris.

The great challenge of the day occurred several years later with the recovery of Comet 1P/Halley. Skillfully trained in the instruments in Paris, Messier anticipated being the first to recover the comet. But he used maps drawn up by his boss, Delisle, that were flawed. Messier searched night after night, on every clear occasion, for 18 months without success. Then, on Christmas night 1758 a German astronomer, Johann Georg Palitzsch (1723–1788), became the first to see Halley's Comet on its return. But news of the recovery spread slowly, and Messier found the comet on January 21, 1759, without having heard of the German observations. By the time the cranky old master Delisle allowed Messier to announce his observation – wanting to allow the German full credit of the rediscovery – the comet was ebbing away in the twilight, some 2 months later.

On Delisle's retirement shortly thereafter, Messier stepped up his observations from the Marine Observatory, which was located in a tall octagonal tower in the Hotel de Cluny in Paris. Soon after his observations of Comet 1P/Halley, Messier discovered his first comet, C/1760 B1. Over the following 25 years he discovered another dozen comets, meticulously recording his observations and analyzing the motions of behavior of the comets he regularly watched.

Messier developed careful methods of painstakingly observing comets. But as most amateur astronomers know, that's not all he did. The fame generated by his comet discoveries made Messier a formidable astronomer in France, despite his lack of academic training and reputation as a "mere" observer. In 1758 Messier became a member of the Royal Society in England, and he aspired to become a part of the French Academy of Sciences. After the death of Nicolas Louis de Lacaille at the early age of 48 in 1762, many scientists considered Messier to be the most celebrated astronomer in France. After his admission into scientific societies in Berlin and St. Petersburg, France finally admitted him as a member of the Academy of Science in 1770.

Messier's admission into the society meant that he could contribute to its journals. Not only was Messier greatly interested in observing comets, but he was repeatedly irritated by running across objects in his eyepieces that appeared to be comets but did not move relative to the stars – fixed objects that could be repeatedly confused with comet discoveries. So he set about creating a list of these celestial nuisances and published it in the *Mémoires de l'Académie* for 1771, which was actually printed in 1774. This publication contained a list of 45 objects, from the Crab Nebula (M1) in Taurus to the Pleiades star cluster (M45).

The so-called Messier Catalog became the gold standard for observers of star clusters, nebulae, and galaxies (although the nature of galaxies would not be known until 1923), despite the fact that Messier created it as a list of nuisances to avoid for potential comet discoverers. Soon after its publication, however, Messier found additional fuzzy objects in the sky, and this process continued, eventually producing a supplement to the catalog, increasing the total number of objects to 68, and publishing the result in the almanac *Connaissance des Temps* ("Knowledge of Time") in 1780.

By this time Messier's comet discoveries had made him nearly legendary. Louis XV of France called him the "ferret of comets." Messier's friend and associate Pierre Méchain (1744–1804) was also a prolific comet observer and discoverer of comets and nebulous objects, and his work helped to spark another extension of the catalog, now totaling 103 objects. Much later, other observers and historians deduced additional objects Messier must have seen, and the total shot to 109 or 110,

Figure 8.1. Comet Lemmon (C/2012 F6), a long-period comet discovered in 2012, put on an impressive show for Southern Hemisphere observers during the first weeks of 2013. This image shot on February 19, 2013, shows the comet's ball-like coma and details in its tail and was made with a 20-inch f/4.5 scope, CCD camera, and stacked exposures. Credit: Damian Peach.

although a few objects involve mistakes and one is a double star. In the end, the Messier Catalog is still considered to be the best list of bright deep-sky objects. In 1801, Messier reflected on the famous list. "What caused me to undertake the catalogue," he wrote,

> was the nebula I discovered above the southern horn of Taurus on September 12, 1758, while observing the comet of that year.... This nebula had such a resemblance to a comet, in its form and brightness, that I endeavored to find others, so that astronomers would not confuse these same nebulae with comets just beginning to shine. I observed further with the proper refractors for the search of comets, and this is the purpose I had in forming the catalogue. After me, the celebrated Herschel published a catalogue of 2,000 which he has observed. This unveiling of the sky, made with instruments of great aperture, does not help in a perusal of the sky for faint comets.

Many other discoverers of comets, professional astronomers and amateurs, followed in the wake of the great systematic observer and cataloger Charles Messier. Despite his humble beginnings, another French astronomer, self-taught Jean-Louis Pons (1761–1831), may have been the greatest visual comet discoverer of all time, having claimed 37 comets. On his heels, the English-American astronomer William R. Brooks (1844–1921) discovered 22 comets. The first woman to discover a comet was German-English astronomer Caroline Herschel (1750–1848), sister of William Herschel and aunt of John Herschel, who aided her brother in his observations and found several comets. In the present century, American astronomer Carolyn Shoemaker has found 32 comets in the Palomar Mountain Observatory searches she conducted with her husband, planetary scientist Eugene M. Shoemaker.

Working along with the Shoemakers and making his own comet discoveries visually as well, Canadian astronomer David H. Levy has found 22 comets altogether, including of course the famous Comet Shoemaker-Levy 9 along with Gene and Carolyn, which mesmerized us all when it crashed into Jupiter in 1994. Other comet seekers have also tallied impressive numbers of discoveries and inspire amateur astronomers the world over with their tales of persistence – if you put in enough dedicated hours behind the eyepiece, you may discover your very own comet too!

The names of a select group of modern searchers pop up frequently in the circles of comet observers. They include Australian observer William A. Bradfield, who has discovered 18 comets, and New Zealand observer Rodney Austin (1945–),who found 3 interesting comets. Lowell Observatory astronomer Robert Burnham Jr. (1931–1993), most celebrated for his famous book *Burnham's Celestial Handbook*, found 6 comets. American astronomer Tom Gehrels (1925–2011) discovered a number of comets, as did American astronomer Eleanor Helin (1932–2009). Australian observer Terry Lovejoy (1966–) has discovered a number of comets, as has his countryman Gordon J. Garradd. The champion comet discoverer is Scottish-Australian astronomer Robert H. McNaught (1956–), who has found 75 in the surveys conducted at Siding Spring Observatory.

In the United States, Donald E. Machholz (1952–) is well known for his comet discoveries, as was Leslie C. Peltier (1900–1980). Canadian Rolf Meier has discovered several comets, as has American astronomer James Scotti (1960–). Those who regularly view comets and read the literature of comet discoveries regularly run across many of the same names, and this fact testifies to the addictive qualities of viewing and searching for comets.

One of the great ways amateur astronomers can be involved in the science of comets is to hunt for them and to find them. The majority of the preceding laundry list of names are backyard observers rather than professional astronomers. While it has become somewhat more difficult over the past few years with the advent of various telescopic search networks and telescopes (see LINEAR, NEAT, PANSTARRS, etc.), it's quite possible to discover comets visually, with a small instrument, from an ordinary location.

Hunting for comets requires a real commitment. Some comet discoverers hunted for years without finding their first comet. Some comet hunters have never discovered a comet. But others have enjoyed the fruits of their persistent labors and added to the knowledge base of science as well as giving their discoveries to a grateful world of astronomy enthusiasts eager to share in the excitement of these celestial visitors.

Figure 8.2. In 2005 the bright Comet Machholz (C/2004 Q2) swung past the Pleiades star cluster (M45) in Taurus, creating a memorable field for observers and astroimagers. This view was captured on January 7, 2005, with a 180 mm lens at f/3.4 and a CCD camera. Credit: Michael Jäger and Gerald Rhemann.

Hunting for comets also provides the benefit of teaching you the sky in a systematic way. In this era of go-to telescopes available at almost every level of expense, far too few amateur observers know the sky well anymore. It's a real shame, because that's one of the greatest feelings of satisfaction you can have as an amateur astronomer – going out under the stars once again and returning to see, whether it be with binoculars or a large Dobsonian at high power, some of the same little features you've enjoyed in the past. Patterns of stars, asterisms, lines of stars, circular features, star clusters – all the little quirky patterns and features you see here and there across the celestial sphere will always be there for you. Comet hunting gives you one of the best ways to learn the sky because it allows you to sweep across large areas of it over and over again.

The hunt for comets is not easy. It's an equal mixture of systematic, repeated behavior, viewing in the right areas of sky at the right times, and a random walk through the chaos of the universe – stumbling onto something no one has ever seen before. The most important part of the equation, just like exercising, eating intelligently, or getting enough sleep, is the commitment. Dedication to the task at hand is paramount. Obviously you're limited to nights that are free of clouds and of other commitments. But to be a comet hunter means to dedicate yourself to a block of time to scan the skies for comets whenever you realistically can, and it's a somewhat rare individual who can keep a commitment like this going, momentum unchecked, for months and years at a time.

Why become a comet hunter? Clearly, you want to find a comet, give that gift to the world, and immortalize your name with the discovery of one of astronomy's rarest and most amazing treats. But you should also want to do it so you can learn the sky, train your eye to see minute details, and become a better overall observer. You should want to so you can immerse yourself in the deep relaxation that accompanies being out under the stars, sweeping your telescope's field of view along, and feeling a close bond with nature. Psychologists call it "self-actualization." Amateur astronomers just think it's fun. But that feeling of being uniquely aware of nature is innately tied to our inherent desire to understand the immensity of the cosmos, to comprehend a little bit of why we're here and how we got here. That's something that slow, careful, patient comet hunting will give you. Even if you don't find a comet that night – or maybe even ever – that seems to be a pretty good deal.

As a prospective comet hunter, you need to understand that you very well may not discover a comet on any given night. The odds are that many hours – perhaps thousands – will pass before you find a comet, or you may never be the first to find one at all. But comets are not trying to trick you. They are still trying to give you one of the best gifts you can receive, because you will discover many interesting things along the way.

In one summer session of sweeping around the sky a few years ago, I noted what I randomly passed, what I "caught" in my telescope's field of view, completely accidentally. I was running my scope along what turned out to be the "spine" of Cygnus and encountered the open cluster NGC 6800, near a bright double star; the nebulous cluster NGC 6823; planetary nebula NGC 6842; the Crescent Nebula (NGC 6888); and various components of faint nebulosity surrounding the bright star Gamma Cygni. That was a pretty good and amazingly visually interesting haul, despite the fact that I saw no comet.

Some successful comet discoverers have searched for 1,000 hours or more before they made their first discovery. So certainly the ability to hang in there and stick with it is important. But luck has something to do with it too – the old saying "sometimes it's better to be lucky than good" certainly plays into comet hunting. Being in the right place at the right time is key. Recall that both Alan Hale and Thomas Bopp were observing the globular star cluster M70 in Sagittarius when they each stumbled onto the 11th-magnitude comet nearby that became Hale-Bopp (C/1995 O1). In 1975, amateur astronomers Doug Berger and Dennis Milon were viewing another globular, M2 in Aquarius, when they became the second and third discoverers of Comet Kobayashi-Berger-Milon (C/1975 N1).

You may have noted that deep-sky targets and comets go together. This is really the case because in small telescopes deep-sky objects tend to look somewhat like comets, especially fuzzy objects like nebulae and galaxies. It's possible to learn to see features in comets very well by training your eye to see features in deep-sky objects. The kind of patient viewing, some of the techniques involved, and the knowledge of how nebulous objects appear will all come from viewing deep-sky objects, even when a good comet is not available.

Some of the education that is gained by viewing nebulae and galaxies will transfer straight across to comets, and in some ways you'll see comets differently than galaxies and nebulae. But the bottom line is that to become an experienced sky observer, you need to hunt down and observe a wide range of objects, both cometary and deep-sky, and use the acuity of your eyes to its maximum.

You can look for comets anywhere, but finding relatively bright comets means looking generally within 90° of the Sun. This concentrates your comet hunting for brighter objects to the early evening and early morning slices of sky – not necessarily in twilight, but not deep in the night at the point in the sky opposite the Sun, either. As far as instrumentation goes, any telescope will do, but a wide-field scope will help you make the most of searching for comets. The wide fields of view from such instruments maximize the contrast between the sky and faint objects, making them pop out a little more effectively, and they also mean you will have fewer fields of view to look into to cover large areas of sky.

Successful comet hunters usually recommend altazimuth mounts for telescopes, allowing simple up-and-down and back-and-forth movements, rather than equatorially mounted instruments. The altazimuth telescope mount will allow you to make left-right sweeping motions and to move up and down easily, covering areas of sky right down to the horizon.

As far as the telescope itself, you should try to outfit it so that it has a minimum field of view of about three-quarters of a degree. Reflectors rather than refractors are best suited for comet sweeping, although some comet hunters have used binoculars as their instrument of choice. The ideal size of the telescope depends on a variety of factors. The sky darkness at your observing site is also involved; the darker the site, the larger the scope you can get away with. But portability is a concern unless you have a permanently mounted observatory scope. The larger the scope you have, the brighter the images will be and the fainter the comets you will be able to see and perhaps to discover. A 6-inch telescope will easily reveal comets that shine at magnitude 8 or brighter. A 12-inch scope will push the brightness down considerably, allowing you to pick up 12th-magnitude comets pretty handily.

Comet hunters also recommend following an organized search pattern. They suggest searching small areas of sky thoroughly, sweeping the sky carefully and slowly as you look through the eyepiece, and moving back and forth to cover one edge of the field of view with the last area you searched – in other words, leaving no gaps. To maximize your chances of success, search in areas that are within 90° of the Sun (before dawn and after dusk), along the place of the ecliptic – the imaginary line that cuts across the sky along which the planets and other solar system bodies appear to travel.

How, exactly, do comet hunters hunt for comets? Any which way that works for you is the best way to do it. David Levy uses an up-and-down pattern of hunting because his telescope moves most easily in that manner. He also favors "hot" areas of the sky that are most likely to produce results first. In other words, he searches in the highest-probability areas within 90° of the Sun first, hitting the evening sky, and later the rich area of the morning sky before dawn. The morning sky in a strip relatively close to the bright dawn twilight represents an area that has been out of view for some time and so can contain comets that have "snuck up" on us from behind the Sun. The farther you stray from the Sun, the less likely you are to find a bright comet, and the more competition you will face with the computerized, wide-field surveys.

The best times to hunt for comets are those following the Full Moon, as greater hours of darkness sweep over the sky with each passing night. Comet discoverers suggest starting to search in the areas closest to the Sun on those nights. Then, until the Moon rises, you have the ability to view a dark sky that has been hidden by moonlight for several nights previously. Look in the evening sky about 2 days after Full Moon, when darkness will reveal a new strip of sky that could conceal a new comet. Wait until the sky is dark enough to allow seeing stars of about 4th magnitude, and stop hunting when the Moon rises.

If you wish to look for new comets in the very fertile period of morning sky prior to dawn, concentrate on periods when the shrinking crescent Moon is thin enough so that it doesn't interfere with overall sky darkness. Search the morning sky 2 hours before dawn, a few days prior to and then continuing a couple days after New Moon. This will give you a fresh piece of sky that may hide as-yet undiscovered comets.

Comet discoveries have not been limited to the nighttime sky, believe it or not. It's possible to discover (and to observe) bright comets when the Sun is above the horizon. But the utmost care must be taken. Never look through binoculars or a telescope in the direction of the Sun. And never look at the Sun with the naked eye, for that matter. Permanent damage can result from ultraviolet radiation in sunlight within moments. The only time to look at the Sun safely without a proper solar filter is during the fleeting moments of totality during a solar eclipse.

In 1985, California comet hunter Donald Machholz produced an unusual statistical study of the 33 comets discovered by amateur astronomers during the decade starting in early 1975. Machholz's paper provides an interesting snapshot of the importance and the variety of discoveries by amateur astronomers, and what they mean.

1. The 33 comets represent a significant minority of the 162 comets discovered or recovered during the period.
2. During this decade, amateur astronomers found an average of 3.3 comets each year, out of an annual average of 16.2 new or returning comets.
3. Machholz surveyed the discoveries made by professional astronomers and concluded that "amateur astronomers would have found perhaps

five more comets if the professional astronomers stopped discovering all comets." The comets not found by amateurs included Comet West (C/1975 V1), Comet 161P/Hartley-IRAS, and Comet Shoemaker (C/1984 U2).

4. Comets are not found regularly. The shortest interval between discoveries was negligible, as Comets 33P/Daniel and 115P/Maury were discovered at nearly the same moment in 1985. The longest intervals between two discoveries were two periods of about 18 months each.
5. A large percentage of comet finds in the evening sky took place between 3 and 7 days past Full Moon. Another large group occurred just before New Moon. The morning discoveries were distributed pretty evenly, from just before Last Quarter Moon to 3 days prior to the next successive Full Moon. Discoveries peaked around First Quarter Moon.
6. The average morning discovery occurred a little more than 30 minutes before the beginning of astronomical twilight, the period when the sky begins to brighten noticeably. The average evening discovery time was about 75 minutes after the end of astronomical twilight.
7. The average brightness of a comet discovered in the evening sky was magnitude 10.2, faint when compared to the average brightness of magnitude 8.5 for discoveries in the morning sky. To explain this discrepancy, Machholz suggested fewer comet hunters look at the morning sky, and comets brighten further before they are found there.
8. Machholz also studied the altitudes of the average comet discoveries in both the morning and evening skies. These figures varied considerably, but he found that evening finds averaged slightly higher than 24° altitude at discovery, while morning discoveries averaged just above 28° in altitude. Machholz suggested that greater numbers of discoveries took place when comets had risen higher than murky hazes along the horizon and above some local obstructions various observers had in their frame of reference.
9. The amateur astronomers who discovered these 33 comets were all men, 26 of them, who were active hobbyists with a driving interest in comets. The champion over the period was Australian William Bradfield, who discovered 10 comets during the decade. Second was Canadian Rolf Meier, with 4 discoveries. Japanese observer Shigehisa Fujikawa had 3 discoveries, and 5 observers each had 2 finds.
10. The average time each observer put into comet hunting, per find, was nearly 282 hours. But the variation in hunting time was very large. Japanese observer Hiroaki Mori spent only an hour sweeping before finding his second comet, and Machholz himself spent 1,700 hours before finding his first comet.

Figure 8.3. Comet PANSTARRS (C/2011 L4) hangs in the sky along with an iridium flare (right) on the evening of February 11, 2013, as seen from Mercedes, Argentina. The imager used a 135 mm lens and a 10-second exposure at ISO 6400. Credit: Luis Argerich.

11. The telescopes used ranged widely in size and type. Visual discoveries took place with a 102 mm (4-inch) reflector (once), 145 mm (5.8-inch) reflector (once), 152 mm (6-inch) reflector (four times), 157 mm (6.2-inch) reflector (once), 202 mm (8-inch) reflector (three times), 254 mm (10-inch) reflector (twice), 404 mm (16-inch) reflector (five times), and 483 mm (19-inch) reflector (once). Refractors were used for comet discoveries of the following apertures: 82 mm (3.3-inch) once, 127 mm (5-inch) once, and 152 mm (6-inch) 11 times. One discovery was made with a 202 mm (8-inch) Schmidt-Cassegrain scope.

In the world of amateur astronomy, comet hunting will definitely delight you whether you do or don't find your own comet. You'll learn the sky in a superlative way, train your eye to see magnificent details in astronomical objects that you wouldn't otherwise have seen, and spend a great amount of time relaxing under the stars, getting away from the hubbub and distractions of everyday life. Those advantages alone ought to be enough to convince you that comet hunting is a seriously worthwhile pursuit.

Let's explore the other possibility – what do you do if you think you've found a comet? The first thing is to check a detailed star chart carefully to see whether there's any deep-sky object – anything "fuzzy" that could be confused as a comet – at the location you're viewing. A nebula? A galaxy? A distant star cluster? A patch of faint stars forming a small asterism? Exactly what can look cometlike depends on a whole slew of factors, including the size and type of your telescope, the sky darkness, atmospheric steadiness ("seeing") and transparency, the magnification you're using, and so on. A few tricks of the trade will help you eliminate some possible mixups.

First, check to see whether your suspected comet seems to show a tail, or whether it's merely a round ball of hazy light. If you gently move the scope's field of view while you look at the suspected comet, you'll eliminate confusion such as reflections from a bright star outside the field, and you'll also employ so-called averted vision, using the rods on the edges of your eye rather than the centered cones. Rods are more sensitive to faint light, while cones are ideal for detecting colors and bright light. The rods will let you see details in the suspected comet better if you gently move it around, slowly, in the field of view.

If you still believe you may have found a comet, try increasing the magnification to higher powers. This will unmask any small clumps of stars that could be masquerading as a nebulous object. If you still suspect a comet, look carefully through the telescope and your finder telescope and record the object's position as carefully as you can, along with your location and the time. Check a detailed star chart to make sure you are not looking at one of the approximately 10,000 deep-sky objects that are relatively readily visible in backyard telescopes. (Now you can appreciate the frustrations Charles Messier must have felt.)

If you still believe you have a comet, make a sketch of the field of view, showing the comet as a realistic, nebulous patch, and plot the positions and relative brightnesses of the stars in the field. Next, look at the same field of view through your telescope half an hour or an hour later. Has the nebulous object moved relative to the stars in the field? If so, you may have discovered a comet. If the object has not moved, you still may have a comet – but don't panic yet. Check a Web site and make sure you're not observing a comet that's already known. Seiichi Yoshida's "Weekly Information about Bright Comets" is an outstanding reference for currently visible comets – its Web address is http://www.aerith.net/comet/weekly/current.html.

Next, try to have your sighting confirmed by a knowledgeable astronomer or expert amateur observer. If you are still convinced you've identified a comet that no one else yet knows about, then you're ready to go on record. The IAU Central Bureau for Astronomical Telegrams in Cambridge, Massachusetts, is the proper authority to which you report a new comet discovery. You can find out how to report a suspected discovery and what information is required here: http://www.cbat.eps.harvard.edu/HowToReportDiscovery.html.

Figure 8.4. A brilliant coma and dusty tail characterized Comet SWAN (C/2006 M4) when this image was made on September 30, 2006, using a 10-inch f/1.5 astrograph, a CCD camera, and stacked exposures. Credit: Michael Jäger and Gerald Rhemann.

Well, if you're like most of us, you love to observe comets but do not intend to make a serious go of discovering comets yourself. If that's the case, there's still plenty for you to know, to practice when you observe comets (and even practice on deep-sky objects), and to learn from the standpoint of observing techniques. Let's say you simply want to observe comets for the fun of it. You'll want to check out the comets that are visible each year with your telescope and compare the appearances of different comets, which do vary considerably. Even the same comet appears quite different from observer to observer, with varying ranges of experience, with different telescopes, magnifications, amount of moonlight presence, seeing and transparency, and so on.

The most basic piece of information comet observers concern themselves with is brightness. Estimating an object's brightness is called making a photometric observation, although for the most part backyard observers are using their eyes to estimate cometary brightness rather than instruments like photometers. The major portions of a comet you should be interested in knowing the brightness of are the nuclear condensation, the coma, and the tail.

It's easiest to estimate the brightness of the coma, especially if the coma is not terrifically large. (Recall that some comae have grown as large as the Full Moon, making estimating their brightness accurately difficult.) It's quite difficult to estimate the brightness of a comet's tail, simply because it is typically so spread out and often in more than one "piece." It can also be difficult to estimate the magnitude of a comet's nuclear condensation accurately because often it's hard to see a small, starlike center of brightness within the central glow. Despite the challenges, it's possible to estimate the brightness of the coma and try to do so with the nuclear condensation, and a consistent range of these estimates from the same observer, sky, and telescope can be very helpful to those studying comets.

The tried and true methods of estimating a comet's brightness originated with English amateur astronomer John B. Sidgwick (1835–1927), American astronomer Nicholas Bobrovnikoff (1896–1988), and German astronomer Max Beyer (1894–1982). American astronomer Charles S. Morris studied these methods extensively around 1980 and produced a new and improved method for estimating cometary magnitudes. The techniques involve "memorizing" the brightness of the comet's image – emblazoning that very clearly into your mind – and then moving to stars that can be compared to the comet's brightness. You should observe the comet with 4× to 5× magnification per inch of telescopic aperture, say 40× magnification for an 8-inch telescope, and then:

1. Defocus the telescope's image until the comet and the comparison stars have a similar apparent size. Note that stars will vary wildly in size when they are defocused but the comet will not.
2. Next, move between a slightly brighter and a slightly fainter comparison star to estimate the comet's brightness in between. With successive pairs of stars, repeat this step until you find a star that matches the comet's brightness. Next, use the average of the measurements as the comet's magnitude, and estimate it if possible to the nearest tenth of a magnitude. You'll need to check the magnitudes of stars with a variety of references, star catalogs, to determine their actual brightnesses.
3. When you've matched the defocused comet and defocused stars, you are comparing to so-called surface brightness of the comet, its brightness spread over a large area, and affixing a magnitude to it relative to the comparison stars.

One of the best sources for comparison magnitudes can be obtained with the *AAVSO Variable Star Atlas* and individual variable star observing charts available from the American Association of Variable Star Observers. Variable star charts can now be plotted online at www.aavso.org/vsp.

Another aspect of cometary observing is measuring the size of a comet's coma. It's very intriguing to know, from the coma's apparent size as we see it on the sky, coupled with the comet's known distance (from an ephemeris), a comet's physical size. So carefully observing and measuring the coma size can be an interesting and important thing to do. The coma size is also helpful in understanding how to make magnitude estimates.

Estimating the size of a coma can be done in an approximate way simply by knowing the true field of view of your telescope's eyepiece and measuring the coma as a fraction of that known distance. You can also make a drawing of the coma and compare that to stars of known separation on an atlas, thereby calculating the size in a relative sense. These are pretty low-accuracy methods, but if done carefully they can produce a reasonable estimate of a coma's diameter.

Another important consideration is knowing the degree of condensation of a coma – in other words, the profile of the light intensity across its face. The degree of condensation ranges from 0 (a diffuse image with no central condensation) to 9 (a starlike, intensely bright condensation at the center). Estimating the degree of condensation will quite simply come with experience, with seeing a variety of comets of different types so that you have an accurate range in mind. Sometimes comets develop a coma that shows a sharp falloff in brightness from a sort of central disk; this unusual appearance rates a 9 on the scale.

Describing the visual attributes of a comet's tail – both ion and dust tails – is something of an acquired art. If a comet has a tail shorter than 10°, then you can pretty well estimate it using the two-star comparison method. If it's longer than that, you're going to have great difficulty accurately estimating its brightness, as so many factors are involved in trying to "compress" such a huge amount of illuminated sky down to the equivalent magnitude of a star. It's best to leave such measurements to electronic photometers, which can measure large areas of sky brightness accurately and obtain reliable total magnitudes for comets.

The next chapter will describe the varied potential for capturing comets with cameras. However, visual observers may also want to record their images of comets on paper by making eyepiece drawings. Capturing comets on paper gives you a permanent record of what you've seen and allows you to compare the appearances of comets on different nights. It's really much easier than you might think.

Sketching at the telescope is quite easy. Set up your telescope some night when a comet (or nebula or galaxy for practice) is visible, and take with you a red flashlight and some sheets of smooth white paper that have comfortably large – say 3-inch – circles drawn on them in black ink. Once you've found a familiar object, choose the eyepiece that shows the object best, get your pencil ready, and start to draw what you see. Sketch lightly. It's easy to get ambitious and rub lots of graphite on paper, smear it around, and come away with a black, shiny mess. Be careful with proportions. First draw in the brightest stars – especially those near the center of the field and ones near the edge – and then carefully fill in star patterns so that all the stars in the field are represented on paper.

As you draw stars, make sure their magnitudes are fairly accurately represented. Make the faintest stars tiny dots and the brighter ones slightly larger. It'll save a good sketch from ruin if you don't draw the stars involved in nebulous light – comet or galaxy – until after drawing the nebulosity. That way you won't smear pencil all over the stars. After plotting the field stars, take a black art pencil like an Eberhard Faber "Ebony" pencil and rub some graphite down onto the paper where the nebulous glow is in the field. You can then use your finger to smear it around a little to create gray tones and the illusion of nebulous light. Where you want to darken the nebulous light, wet your finger slightly and rub a little more. If you have to, use an eraser to shape the nebulous glow or to correct mistakes.

You don't have to make a finished sketch at the eyepiece. Make a rough drawing outside and then take your time later indoors to finish and clean up the sketch. If you do this, however, you must be honest: Become familiar with the object and take everything down in rough form at the telescope; otherwise, you might exaggerate the view and come away with a phony sketch.

Drawings can be a huge help in studying fine details in comets that images may not even capture. These include halos, circular envelopes of light surrounding the central condensation. You may also note and draw fans, which are sectors of bright material originating from the central condensation and moving away from it in a radial pattern. Notable features may also include rays of jets, sharply defined areas of detail arcing away from the central condensation. You may also see envelopes, several layers of brightness immersed in one another, as you move farther away from the nucleus. You may also note spines, sharp streaks connecting the central condensation and various elements of the tail. Curiously, you may also see what appears to be a shadow of the nucleus, which is really a dark-appearing streak leading from the central condensation toward the tail. And finally, you might spot streamers, nebulous, bright streaks in the coma or tail.

When you make a comet drawing, carefully consider all these phenomena, and observe the comet at a range of magnifications to see which allows recording the comet in the best way. You'll see that averted vision, looking slightly to the edge of the field to utilize the rods in your eyes, will help with seeing faint details. Make you sure you're well dark adapted before making drawings; your eyes will be far more sensitive to faint light after being in the darkness for an hour, as opposed to right after going outside from a brightly lit room.

Observe the comet carefully before you start making a drawing, and consider all the details at length. Draw the brightest stars first and then fainter stars, and then add graphite to represent the comet's center and coma. Finally, you can add details such as those described. Make sure you carefully consider aspects of the details; draw the correct brightnesses, orientations, and spatial relationships as best you can. The easiest way is to draw with pencil on white paper, creating a "negative" drawing. You can also draw with various white pencils on black art paper for a "positive" drawing, but these supplies do not allow recording details as finely as you can with negative drawings.

You can use any telescope or even binoculars to make comet renderings. But since some of the fine details in a comet are best visible at high resolution and with high magnifications, perhaps the best instruments are long-focus, apochromatic

Figure 8.5. Comet 62P/Tsuchinshan appears to be an extra galaxy in the field of the center of the Virgo Cluster of galaxies, photographed on January 16, 2005. The bright elliptical galaxy at center is M86, and the comet is the delicate, oval fuzzball just to its right. To the lower right of the comet is another bright elliptical galaxy, M84. Edge-on galaxies lie above (NGC 4402) and below (NGC 4388) of the comet, while the interacting galaxy pair NGC 4435/38 lies to the left. The imager used a 200/300 mm Schmidt telescope, a CCD camera, and stacked exposures. Credit: Michael Jäger and Gerald Rhemann.

refractors. Make sure you carefully note all important data with each drawing, which include the comet, observer, telescope, aperture, focal ratio, magnification, date, and time. You can also produce a short written summary of any interesting features you have observed.

In an age of high-resolution digital imaging, you might wonder about why drawings are important at all. They not only capture comets as humans see them – images often overexpose bright parts of sky objects and capture other parts not visible to the eye – but they allow making the most of one great advantage the human eye has. Eyes have an enormous dynamic range, being able to see an incredible range of brightnesses and levels of contrast. Because of this, the eye can capture fine details in a comet that a camera might not record. This is especially true for moments of good seeing, when the atmosphere steadies to become exceptionally smooth, so keep looking at details in comets and you'll see these periods of high crispness in detail.

Even professional astronomers value high-quality drawings in this CCD age. Capturing an accurate state of the coma and nuclear condensation can be valuable in determining the rotation rates of comets. This is so because high-resolution drawings can include areas of "hot spots" where cometary nuclei are outgassing and where in the inner coma concentrations of light indicate dense material. Few amateur astronomers make high-caliber drawings of comets, but it's an area of real value and creates a record of how particular comets change during any given apparition.

When it comes to capturing souvenirs of comets, most amateur astronomers are focused on the activity that used to be much harder than it is now. Over the past 20 years, astrophotography, now called astroimaging, has made huge strides with advances in technology that make it easier than ever to capture comets in digital images that will last forever.

9

Imaging Comets

A generation ago, capturing your own photographic portraits of the sky was a difficult and complex process. Cameras could be tricky, choosing the right film was an arduous process, and technology simply wasn't what it is today – the process of guiding your telescope accurately as Earth rotated and the sky seemed to move could be such an exacting process that it drove people crazy. Now, producing high-quality pictures of the sky is still a somewhat tricky process, but it's much easier than it used to be.

As with any sky targets, you can capture pictures of comets in a variety of ways. Astronomical objects are very faint compared with normal scenes of people at the beach, so long exposures are generally needed, regardless of the equipment setup or technique used.

Types of comet photography range from pretty simple to quite complex and of course demand increasing sophistication of equipment and knowledge. The easiest way to capture a comet is simply to mount a camera on a tripod, fitted with a wide-field lens, and shoot an unguided exposure that will capture the comet amid constellations. More ambitious is to mount your camera "piggyback" style on a guided platform or telescope, to capture a time exposure of a smaller field of view that tracks the sky to preserve small, round star images. The most sophisticated method is prime focus photography, in which the camera is attached to the telescope and the telescope is used as a giant lens. Because of the high resolution and small field of view, the tracking must be superb and the exposure calculated to capture the desired details in the comet. All three basic methods will be discussed in this chapter.

Another aspect of astrophotography has transformed radically over the past generation. Twenty years ago amateur astronomers mostly used film and developed it to produce slides or prints, either commercially or with their own home darkrooms, to maximize the quality of images. Now amateur astronomers use digital single-lens reflex (DSLR) or even dedicated CCD cameras – digital cameras designed for astroimaging – and the product is an electronic image, stored as a computer file and consisting of pixels of varying brightnesses and colors. The "darkroom" now consists of Adobe Photoshop or another imaging processing software program. This makes the processing of images much less messy and tremendously faster, but it also raises questions about processing images correctly and not overprocessing them.

The first step in entering the world of astrophotography is to choose a camera that will do what you want it to do. The range of cameras to choose from is now greater than ever, making the right choice a little trickier than it used to be. All cameras now use electronic image sensors. The sensors in cameras now consist of one of two basic types: charge-coupled devices (CCDs) or complementary metal-oxide semiconductors (CMOS), each of which offers a viable method for producing good astrophotos. Each of these types of detectors uses an array of sensitive cells that produce picture elements (pixels) to make up the image. The greater the number of pixels, the higher the resolution of the image. The pixel number can range from 640 by 480 in small images of planets to millions of pixels in images of galaxies and nebulae.

Before you choose a camera for astroimaging, consider what you're buying carefully. Do you want to be able to use the camera for normal, everyday pictures at the beach, as well as for astronomy? If so, you need to buy a DSLR. If you're satisfied with buying a camera for astroimaging only, then other types may work well for you.

Some image sensors are made for capturing color images and others are made simply for black and white work. Monochromatic sensors produce sharper images, and you can use them to create color images by shooting three filtered exposures in red, green, and blue, then combining them digitally. Straight color sensors use a matrix that alternates filtered pixels in different colors to create a single, overall color picture. But it will not be as sharp as an image made with the monochromatic sensors.

If you choose a DSLR camera, you'll get a cost-effective camera that can be used for normal photography as well as sky shooting. They are great for shooting comets and deep-sky objects and can be easily piggybacked onto a telescope and used for guided, wide-field imaging as well as being connected to a telescope at prime focus and used for through-the-scope, close-up photography. If you want to use a dedicated CCD camera for astroimaging, you'll also need to use a laptop computer with the telescope to control it.

The camera brand will have a big impact on your astroimaging. While many brands have been used for successful astroimaging in the past, Canon is the only major manufacturer to produce a camera specifically made for astrophotography, their EOS 60Da model introduced in 2012, and their other models include some features that are specifically helpful for astroimaging.

Features you'll need with a DSLR include live focusing, which means you can focus the camera while you look at images in its screen, with a magnified portion of

Figure 9.1. Comet 73P/Schwassmann-Wachmann (Schwassmann-Wachmann 3) is a periodic comet that reaches perihelion every 5.4 years and is currently in the process of disintegrating. This remarkable image shows two of the fragments traveling in parallel, each with a distinct tail, and was made May 2, 2006 using a 300 mm lens at f/4, a CCD camera, and stacked exposures. Credit: Gerald Rhemann and Michael Jäger.

the image falling onto the sensor. This allows you to focus sharply on star images, which need to be precise, small pinpoints. Autofocusing is not applicable to astronomical imaging. You'll also need a cable release, which allows you to start and stop exposures without bumping the camera, and extra batteries.

Choosing the right lenses is also important. The zoom lenses that are usually sold with camera bodies are not well suited to astroimaging; they have apertures too small to be useful and they're not terribly sharp. The best thing to do for wide-field imaging is to get a 50 mm f/1.8 lens, which will be photographically fast and sharp. You may want even wider lenses to shoot "all-sky" images or, more likely, telephoto lenses for piggybacking your camera on a guided telescope. Telephoto lenses exist in a wide array of focal lengths. A 200 mm lens is a good middle of the road telephoto that will produce much closer images of comets than the 50 mm lens.

The other main option, aside from a DSLR, is to get a dedicated CCD camera made specifically for astronomical imaging. A wide variety of makes and models exist, including those manufactured by Santa Barbara Imaging Group (SBIG), Quantum Scientific Imaging (QSI), Imaging Source, Finger Lakes, Starlight Xpress, Apogee Instruments, Atik, Celestron, Orion, Meade, and others. Long used by dedicated deep-sky imagers, these sophisticated instruments are now more affordable than ever before and offer high-resolution sensors that make imaging objects like comets at high resolution simpler than in years past.

Dedicated CCD cameras require a computer to "run" the camera/telescope combination. Your laptop will need to be red filtered or display its screen in a "night vision" mode so your dark adaptation will not be ruined. CCD cameras are cooled to maximize the sensitivity of the collecting array, and these images represent the state of the art in terms of quality. But the work involved is significantly further beyond that which can be done with a simple DSLR.

In recent years, some astroimagers have been producing satisfying results with video imaging. Using a video camera to capture solar system objects allows choosing the best moments of atmospheric seeing, letting the computer select the sharpest frames to combine into a final image. Originating with inexpensive webcams, astronomical video cameras now include the MallinCam, Orion StarShoot Deep Space Video Camera, Stellacam, and others. Color cameras and monochrome cameras exist, as do video astrocameras that can also function as autoguiders for your telescopic long exposures. These systems also require a computer connection and control.

What about simple, non-DSLR digital cameras? These are ubiquitous, in every hand of every tourist, but can they accomplish anything when pointed skyward? Point-and-shoot models, though many have large megapixel counts, have smaller sensors that are not well suited to imaging faint objects. They can be used for afocal photography – that is, simply pointing the camera's lens into a telescope's eyepiece and shooting away. But generally they are not particularly good for astroimaging.

Once you've chosen a camera, consider taking some examples of the simplest kind of astroimaging, camera-and-tripod photography. This technique consists of simply mounting a camera on a fixed tripod and making relatively short exposures that are not guided to compensate for Earth's rotation. Thus, wide-angle lenses are needed. The standard focal length of 50 mm will work fine, but you may want a wider angle of something like a 28 mm lens, or even a fisheye that will capture the whole sky with an 8 mm–16 mm focal length. Fisheye lenses are expensive, however.

Choose a tripod that will not create problems. Many photographic tripods look nice but are weakly built, providing a wobbly platform, particularly when they're fully extended or when they're outside with something of a slight breeze blowing. I recommend getting a tripod that is overbuilt – and the Tiltall that I use is more than substantial enough.

Once you head out into the field, you are ready to try your first comet shots. Or, if you're waiting for a bright comet that hasn't yet appeared, you can practice camera-and-tripod shots on Milky Way star fields, aurorae, planetary conjunctions, or simply constellations. You need to keep some basic rules in mind, however.

Composing a good comet shot requires you to think like a professional photographer. A good compositional rule to keep in mind is the so-called rule of thirds, to position the main subject – the comet – in the right place. Imagine your camera's field of view as a frame divided into thirds, both horizontally and vertically. This slices it into areas that are equivalent to a tic-tac-toe board. If you place the comet near one of the four points that represent intersections of the lines on that board, rather than always dead-centering it, your photo will have maximum dramatic appeal.

The horizon will look best if you place it in the bottom third of the picture. The off-centering of the major elements like the horizon and the comet itself will help build drama by creating a slight tension in the perceived scene, rather than a static, perfectly balanced image devoid of drama.

You will also have to travel to the best possible site you can reach to photograph the comet. If you live in a city or even in the suburbs, your sky may not be dark enough for satisfactory photography of comets, even if they're bright. Be ready to travel to at least a reasonably dark sky site and scout it in advance to make sure the horizons are free of obstructions. Light pollution can be the major factor that limits astronomy enthusiasts in what they can accomplish, and it's never a bigger deal than with faint objects like comets and nebulous deep-sky objects.

Before you head out for a photographic session, you'll want to create or plot finder charts that help you locate the comet easily and choose bright stars for composition and for guiding if necessary. For wide-field, fixed photography, however, at least the equipment questions are simple.

You'll want to arrive at the observing site before dusk or, in the morning sky, long before the comet is ready to be photographed. Check out the composition of the scene in front of you, including foreground Earth that will end up in the picture. A few steps in one direction or another can make a big difference in the resulting photo. Once you have a good spot from which to photograph, simply look through the camera's viewfinder at eye level. If you choose to shoot a horizontal picture, make sure you check the vertical composition to see whether it looks better.

Next, set up your tripod, mount the camera on it, and attach the accessories. Use the best lens you can for the situation and the composition. Check the parts of the scene that will end up in the photo – do you have the horizon, the comet, the bright stars, and so forth, in the right place? If not, make the necessary adjustments.

You can add drama to wide-field shots by including silhouettes in the field of view. An observatory, house, cactus, mountain, or other object can add drama, even if it's only marginally visible as an outline in the finished photo.

Exposure times for wide-field comet shots will vary considerably, so some degree of experimenting is necessary. If you want a picture of the comet with nice, point-like stars, exposures will have to be kept short. Shooting at an effective high speed for 30 seconds or so ought to do that. If you want to push things further, you might get stars – and comets – that appear to trail slightly because of Earth's rotation. So experiment to see what works best.

The next step in comet photography is to piggyback your camera onto a motor-driven telescope that will compensate for Earth's rotation and therefore allow much longer exposures while preserving pointlike stars and good cometary detail. This involves polar aligning your telescope so it's oriented correctly relative to the North Celestial Pole, the point around which the stars appear to circle (or, in the Southern

Figure 9.2. Comet 63P/Wild 1 was a mere fuzzy "star" with a diminutive coma when photographed on February 27, 2013, using a 17-inch scope at f/4.5, a CCD camera, and stacked exposures. Credit: Damian Peach.

Hemisphere, the South Celestial Pole). Then you start an exposure and make subtle corrections in tracking during the duration of the shutter opening – telescope drives are not perfect on their own – or employ a computerized autoguider to make the corrections. This allows getting really fine images of comets that are closer up than camera-and-tripod photography can permit.

In order to accomplish piggyback photography, you need to have a telescope with an equatorial mounting. Altazimuth mounts will not work because of a phenomenon called field rotation. Long exposures with cameras mounted on altazimuth structures produce images of stars near the edges of the field that trail around the center of the field. This is because the orientation of the field of view gradually changes as the telescope tracks along.

Accurately polar aligning a telescope such that it tracks the sky properly is an important step in the process. It becomes more and more critical if you plan to use lenses with relatively long focal lengths, such as longer than the standard 50 mm. The greater you magnify the image, the better the tracking has to be. And a precisely polar aligned telescope means finding objects will be easier, even if you have a go-to computerized mount.

It only takes a few minutes to use your finder scope to achieve polar alignment good enough for shooting with wide-field lenses, up to 50 mm in focal length. First, you need to make sure the finder scope is precisely collimated with your telescope – that is, both point at exactly the same spot. You can do this most easily during the daytime by aiming the scope and finder scope at a very distant, small object – a light, a tree branch, a fence post, for instance – and making sure that the scope is centered on the object, and then precisely centering the finder on the same small feature.

Next, under a dark sky, center a bright star in the finder scope, turn on the telescope's drive, and center the star in the telescope too. Now make sure the star appears at the exact intersection of the finder's crosshairs. And make sure the star is still centered in the telescope. If it isn't, repeat the adjustment process.

Now rotate the finder scope so the crosshairs parallel the declination and right ascension axes of the telescope. You can make sure you've done this properly by finding a bright star and moving the scope in right ascension and declination, seeing the star move right along the crosshairs. Now determine your finder's field of view by turning off the telescope's drive and watching a star near the celestial equator drifting through the finder scope's field, bisecting it. You can then divide the number of minutes the star takes to drift through the field by 4 to calculate the finder's field of view in degrees. (Finder scopes often have fields of 4°.)

You can now approximate the placement of the pole in your finder scope, to establish polar alignment, by knowing that the North Celestial Pole lies 0.7° from Polaris, the brightest star in Ursa Minor, in the direction of the Big Dipper. You may well have a telescope with a built-in polar alignment telescope, or a computerized go-to scope with a built-in routine for finding the pole. If this is the case, consider yourself fortunate.

Once you have the telescope aligned and the drive running, you're ready to produce a test exposure. Good polar alignment, along with a piggybacked, wide-field lens, allow you to expose for an hour or so without much trouble. Again, experimentation is the key to results that please you.

You should always remember the basic rules of astrophotography, however, because when it's late or early, you're in the dark, you are trying to keep track of many things, and you're excited over the appearance of a comet, many things can and will go wrong. Murphy's law will sometimes rule the day. Don't let yourself become discouraged if you go through some rough experiences, especially early on in your astroimaging career.

Make sure the lens cap is removed from the camera. Make sure the camera's shutter is held open with a cable release or other suitable mechanism. Keep your lenses clean. Set the focus and f/stop to the right configurations. Center a bright star in both the telescope's field of view and the camera lens to make sure you have good rough alignment between them. Firmly secure the camera to the telescope with

Figure 9.3. Comet 10P/Tempel 2 floats like a cotton ball against a starry background in this image made July 16, 2010, using an 8-inch f/3.6 astrograph, a CCD camera, and stacked exposures. Credit: Gerald Rhemann.

a locked plate. Set the camera's shutter to the proper "B" or "T" setting for a long exposure. And keep good data for all of the images you make, including date, time, duration of exposure, full equipment data, sky conditions, and anything else that could be of interest later on. You will thank yourself on down the road.

As with any celestial photography, practice makes perfect. Shooting a number of piggybacked astrophotos will familiarize you with the system and the setup, and your images will certainly get better over time. You'll want to practice with deep-sky objects so that when a nice comet appears, you have the procedure down and don't waste opportunities learning subtleties when a prime target is in the sky.

If you want to step up to the big leagues, you can shoot comets directly through your telescope, attaching your camera to the scope and using it as a giant lens. Through-the-scope, or prime focus, photography is a demanding challenge. For example, if you attach your camera to an 8-inch Schmidt-Cassegrain telescope (SCT) with a focal ratio of 10, it becomes in effect a 2,032 mm telephoto lens (80-inch focal length times 25.4 mm per inch). For the sake of argument, assume that a normal 50 mm lens has a magnifying power of 1 – that is, it doesn't magnify or reduce the image relative to the way your eye sees it. By that standard, an 8-inch SCT has a relative magnification of about 41 – a field of view 41 times smaller, and closer up, than those 50 mm shots you were taking previously.

This enables exceptional close-ups of comets, showing tremendous detail in their nuclear condensations, comae, and gas and dust tails. The details you can capture now would be too small to be revealed in piggyback photos – even ones made with long-focus lenses. But remember that every mistake in guiding, gust of wind, or erratic bump to the telescope will now be magnified by a factor of 41. This means that focusing must now be precisely sharp, drive corrections nearly perfect, and the exposure just right to show what you want to show. Errors will now stick out like sore thumbs. Finding the object you want to photograph (if the comets are faint), finding suitable guide stars, and managing other complexities will all be a bit harder now. Polar alignment will be far more critical. Producing superb results now means you will really need to be precise in the way you work each part of the photo process.

To do prime focus imaging, you'll need everything you needed for piggyback photography plus a few additional items. You will need to have an equatorially mounted scope on a tripod or pier. You could use a German equatorial mount or a fork mount. Because of field rotation, altazimuth mountings are out. (You could use one if you have a gadget called a field derotator, but the complexity required for success is challenging.) You will have to polar-align your telescope to a high degree of accuracy.

You will also need the correct type of adapter to connect your camera body to the telescope's prime focus. Most major telescope companies offer a good selection of so-called T-rings and T-adapters that enable connecting a variety of camera brands to their telescopes. Find out which one you need and procure it.

If you are going to guide manually, you will need a good illuminated reticle guiding eyepiece. These are usually eyepieces of 12 mm focal length with a variable brightness LED illumination that lights twin crosshairs in the eyepiece that enable you to watch with precision a guide star and make corrections with your telescope's drive to keep the star precisely centered during the length of the exposure. These typically plug directly into the telescope's power unit or run from a separate battery pack.

You will need the right type of 35 mm camera body. The best types for astroimaging are often "obsolete" types that are mechanically operated and have shutters that can be locked open without draining the battery. Some cameras have been produced with astroimaging in mind, such as the previously mentioned Canon 60Da. Whichever camera you choose, it needs to have a bright focusing screen that will enable you to see dim targets and make possible focusing on relatively dim stars. Some telescope companies offer these as accessories for cameras that have interchangeable focusing screen capability. If you also procure a right-angle finder for the camera, it will save lots of bending of your neck in uncomfortable positions.

You may also want to consider protection against dew, which can make a mess of a nighttime photographic session. This is particularly the case with SCTs and other telescopes that have a front lens on which dew can easily condense in damp climates. The best way to protect yourself against dew is to shoot astrophotos from a dry location. But in many parts of the world, that just can't happen. Taking along a portable hair dryer that can periodically be turned on and gently carried over the surface of the lens is the best way to get rid of dew. A lens hood can also protect

Figure 9.4. Comet PANSTARRS (C/2011 L4) basks in an Arizona sunset (left) along with a young Moon in this image taken March 12, 2013, with a 200 mm lens at ISO 400. Credit: Chris Schur.

your scope against the formation of dew. Some manufacturers offer electrically heated dewcaps, which make a nice accessory for the amateur astronomer who has it all.

Another accessory you'll need is an off-axis guider. Placed between the telescope's rear cell and the camera, such a device contains a small prism that deflects a portion of the light at a right angle into a tube that will host the guiding eyepiece. This borrowing of a portion of the light, most of which continues right on through to the camera, allows you to see a guide star. You can then make guiding corrections with the scope's drive corrector as the exposure runs. An off-axis guider is handy, but it means you need to find a suitably bright star to guide on relatively close to the object you're photographing. The guide stars will appear to be somewhat dimmed because of the use of the prism, which takes a little light away from the main image. In star-poor areas of sky, you may struggle somewhat to find a suitable guide star close enough to the comet or other object you're imaging.

You could also use a separate guide scope, a telescope mounted on top of the scope you're imaging with, that hosts the guiding eyepiece. These are often small refractors of long focus, but 3.5-inch or 5-inch SCTs also work well for this purpose. This gives you a big advantage relative to an off-axis guider in that finding a suitably bright guide star is made much easier. You can find any reasonably bright star in a larger area of sky to use as a guide star in the separate guide scope. You can move the guide scope's mounting a slight distance up and down or left and right – relative to the main scope – to find a suitable star. The separate guide scope does offer a new problem, however, in that during the course of a long exposure, flexure can set in. That is, the guide scope can move in tiny increments differently than the main scope does, and that can introduce small errors in the image. The additional weight from the guide scope will also mean balancing the whole system carefully so that it tracks smoothly.

With the application of a little money and technology, you can avoid all of the hassle of off-axis guiders, peering into an eyepiece, making the guiding corrections, and using illuminated reticle eyepieces. Several manufacturers make automatic guiders that are in essence modified, simple CCD cameras that are equipped to detect the motion of a guide star and instruct the telescope's drive system to move in the appropriate direction to keep the star centered. This extra expense may well be worth it, if you're going to be a dedicated sky photographer. Most serious astroimagers have autoguiders and thank their lucky stars for them, remembering the days of long exposures and painstaking manual corrections as a past horror.

Another consideration for shooting through the scope is a focal reducing lens, sometimes called a telecompressor. Because of the very long focal ratios of many modern scopes, such as f/10, they are photographically "slow" and require somewhat long exposure times. A focal reducing lens in effect cuts the f/ratio of the system, however, typically by two stops – making an f/10 scope effectively f/5. This shortens required exposure times by approximately 75 percent, and although it reduces the effective area of the image, that is not a big problem.

Before you get ready to shoot your first telescopic comet exposures, make a list of your target or targets, and take along the coordinates of the comets. Whether your scope has go-to technology or not, you need to be able to find the comets reliably quickly to set up for your photographic work. The best thing to do is either to input the coordinates of the comet into your telescope's onboard computer, if it has that capability, or to plot the comet's daily positions on a star atlas and take it to the telescope with you.

Once you've centered the comet or at least the comet's known position in your camera's viewfinder, you can acquire a suitable guide star. Do this by finding the nearest star that is bright enough to guide on with an off-axis system, or by finding a star with your separate finder scope in the vicinity, locking onto it, and getting it ready either in your illuminated reticle eyepiece or in your autoguider. You may need to rotate the camera and guider in order to position the guide star appropriately so it can be seen in the guiding eyepiece.

Once you have acquired the guide star, you need to focus the stars in the camera's finder, making sure they are critically sharp. You cannot focus on a comet. They are much too fuzzy to try good focusing on, so always make sure you are focusing to make stars in the field the sharpest pinpoints you can. Travel back and forth on either side of sharpest focus several times to make sure you have the cleanest, most precise focus possible. And you will thank yourself if you make it a regular practice to refocus the telescope after every exposure. Little vibrations, variations in temperature, and other factors can throw the telescope out of focus during a night's photographic session.

If you're guiding manually with an illuminated reticle eyepiece, you can now turn the guiding eyepiece in its tube so any movement of the guide star will be along one of the reticles, making corrections easier. Center the guide star in the crosshairs, start a timing of the exposure, and begin by opening the camera's shutter. Lock the shutter open with the cable release. Using your scope's drive controls, keep the star at the intersection of the crosshairs during the entire exposure. If the star drifts to the east or west, this may be caused by irregularities in the telescope's drive gears. They often are not made to a high enough precision to avoid so-called periodic errors as they track. Some telescopes do feature periodic error correction software in which the scope "learns" its own tracking errors and then applies corrections to them. If the guide star drifts more in a north/south direction, then you probably need to improve the scope's polar alignment.

When your exposure is finished, close the shutter. You have captured the image of a comet and now the fun of image processing can begin.

Alternatively, you may be using a dedicated CCD camera for imaging rather than a traditional DSLR. With a CCD camera, the process is very similar, but a few differences apply. As mentioned, using a CCD camera requires a computer, and one that can be used in the field at the telescope. Some CCD cameras are one-shot color cameras, but most – and those that offer the best resolution – are monochromatic cameras. To make color portraits with them, you take three separate exposures of the same object, one each in red, green, and blue (RGB) filters. An infrared blocking filter is also used. The three exposures are combined digitally in Photoshop or a similar program. The final result is termed an RGB image, as a composite with each of the three primary colors. An equal exposure made with each of the three filters produces a balanced, final color shot. However, a CCD camera has a greater sensitivity to red than to blue, so some magic has to take place to make the shots turn out right.

So the complexities of CCD imaging can fill volumes. For a typical SBIG camera, one of the leading manufacturers, cutting-edge CCD imager Jack Newton, uses a ratio of 1 part red to 5 parts green to 16 parts blue, in terms of exposure times. Thus, CCD imaging of faint astronomical objects means shooting a series of images and combining them all in a program like Photoshop to achieve the final desired result. It's a complex process that requires lots of practice and skill in order to achieve proficiency.

Sky shooters recommend a good 8-inch or 10-inch SCT as a platform for CCD imaging and suggest go-to capability. The addition of a CCD camera means rebalancing the mount with counterweights. This system requires significant cabling, which must be tied down so as not to interfere with the telescope's movements. An altazimuth telescope can be used with a field derotator, which prevents the star field from rotating relative to the CCD camera's field. You will thank yourself, however, if you have an equatorial wedge and align the telescope precisely with the pole. As with any through-the-scope imaging, you'll need a guide system such as a guide scope with an autoguider. The finder scope will need to be aligned precisely with the telescope, as you won't be able to see through the main scope once the CCD camera is attached.

As always, focusing is critical. Imagers recommend selecting a bright star close to the target and taking a 1-second exposure. The star will be unfocused. Change the focus. Take another 1-second exposure. Keep this process going until you move the focus to precise sharpness and keep checking it with the short exposures.

You can now center the comet or other object you're shooting and take a short exposure. Is it positioned correctly in the frame? When you're happy with the composition, you can take a 1-minute exposure. If the stars are nicely round, feel free to take a longer exposure. If you're using an altazimuth mount and want to stack many exposures, a program like *MaxIm DL* can help significantly.

You can capture images with the different filters and combine them in Photoshop or with a whole array of accessory programs. Photoshop has for the majority of astroimagers become the modern darkroom. The complexity of Photoshop can be enormous, and how you'll want to process images depends on a large variety of factors. A number of good books exist on astronomical image processing with Photoshop and other programs, and the best thing to do to wade into the pool of digital astronomical processing is to acquire one of these books. You can also read leading astroimager Tony Hallas's column in *Astronomy* magazine, which every month describes image processing techniques and how to make the most of them.

A few details about astroimaging pertain specifically to comets. Because comets move at a different rate than the stars appear to, you can track on stars or, to produce a more accurate image of the comet itself, track on the comet's nucleus. The most accurate method is rather detailed: It involves calculating the comet's differential motions in right ascension and declination and driving the telescope at those speeds relative to the scope's normal drive motions. Some drive correctors with go-to telescopes now have the capability to drive in variable rates like this.

Another way to do it is simply to guide on the comet's nucleus, its visually brightest condensation within the coma. An autoguider can be instructed to do this, or you can do it manually using an illuminated reticle eyepiece. You can also center the coma in the crosshairs and try to keep the coma centered in the same way throughout the exposure, but this can lead to pretty inaccurate results.

However you capture images of comets, you are sure to learn a lot about astronomy, the sky, and comet science in the process. Even if your images don't turn out to be as magnificent as some of those that are published in the popular magazines,

Figure 9.5. Comet Garradd (C/2009 P1) showed a broad, fanlike dust tail when it was imaged on November 22, 2011, with an 8-inch f/3 hyperbolic astrograph, a CCD camera, and stacked exposures. Credit: Gerald Rhemann.

creating your own comet shots gives you a permanent and personal record of the comets you've experienced, and that makes for a pretty special, lifelong goal.

As Comet ISON approaches us in late 2013 and we hope for another Great Comet, we've enjoyed looking back on the incredible history of bright comets of the past. Recent ones, Ikeya-Seki, West, Hyakutake, and Hale-Bopp, have produced spectacular memories that linger on with amateur astronomy enthusiasts around the world. We've looked back on some of history's greatest comets, the Great Comet of 1577, the Great Comet of 1811, Halley's Comet, and many more. We've seen how comets have been perceived by humans over the millennia, from Aristotle through Roman times, into the murky Dark Ages, and finally out into the Renaissance. We've seen how Copernicus, Galileo, Newton, Halley, Messier, and many others have thought about comets, and how their thoughts have changed ours.

The most recent findings in planetary science, helped by ingenious astronomers, space probes, and meticulous ground-based observations, have unveiled the role of comets in the solar system. They harken back to the earliest days of our star system's formation; hold primitive ices in their great, distant lair part of the way to the next star system; and occasionally thunder in to greet us with a new appearance, giving away their gases and dust and reminding us the solar system is a very dynamic place. We have seen the destructive power of comets and their cousins the asteroids, this past year in Russia, a century ago in the same region, and 66 million years ago in a planet-changing doomsday scenario. We've seen that comets and asteroids can cause mass death but they also hold amino acids and other organics, the cosmic seeds of life.

When you're out under the stars, in evening or early morning, watching a bright comet, there's a very personal attachment you feel to the universe. The everyday problems of life on Earth melt away. You are there, alone in the cosmos, with a very large and amazing presence before you. It's an electric feeling of discovery, of realizing your place in the universe. It's a feeling I first had 35 years ago when gazing up at Comet West. It's a feeling I hope you'll identify with out under the stars with Comet ISON, or other comets, stars, nebulae, and galaxies, time and time again.

When you feel it, you'll know it. And you'll understand that you will never see things quite the same way again.

Glossary

Abiogenesis The natural process by which life arose from inorganic matter.

Amino acids Biologically important compounds made from amine and carboxylic acid groups, and constituting the building blocks of proteins.

Antitail When Earth crosses the orbital plane of a comet, projection effects sometimes cause what appears to be a sunward-pointing component of a comet's tail.

Aphelion The farthest point from the Sun in a solar system object's orbit.

Apogee The farthest point a solar system body reaches from Earth in its orbit.

Apparition A good period of visibility of an object as seen from Earth.

Asteroid Small rocky body in the solar system, also called a minor planet.

Astrometry Positional astronomy that focuses on determining the precise positions of objects in the sky.

Astronomical Unit The average distance between Earth and the Sun, defined as 149.598 million km.

Calcium-aluminum inclusion Inclusion found in carbonaceous chondrite meteorites, rich in aluminum and calcium, that is believed to be the oldest substance in the solar system, having formed about 4.57 billion years ago.

Carbonaceous chondrite A primitive class of meteorites that formed on the outer edge of the solar system and that contains primitive matter.

CCD camera A camera employing a charge-coupled device, an electronic photoreceptor that allows capturing a digital image with light-sensitive pixels (picture elements).

Centaur A class of small solar system body with a semimajor axis between the gas giant planets, characterized by an unstable orbit that crosses the orbits of the giant planets.

Central Bureau for Astronomical Telegrams The official international clearinghouse for information about transient astronomical phenomena, including comets.

CHON particles Particles of primitive matter in comets and asteroids that are rich in carbon, hydrogen, oxygen, and nitrogen and represent primitive matter from the early solar system.

Coma The extended area of gas and dust surrounding a comet's nucleus that is mostly spherical and is not strongly affected by the solar wind or radiation pressure.

Cometary fading Concept created by astronomer Jan H. Oort to explain why the number of comets returning to the inner solar system appears to be less than astronomers predict.

Cretaceous-Paleogene extinction event The K-Pg Extinction (for short, and also called the K-T Impact for Cretaceous-Tertiary, for the now-outmoded term Tertiary) is a global extinction event that occurred about 66 million years ago and was caused by a roughly 10-km asteroid plunging into what is now the Yucatán Peninsula in Mexico, creating Chicxulub Crater.

Cubewano A nickname for a classical Kuiper Belt Object, a low-eccentricity icy small body that orbits beyond Neptune and is not in gravitational resonance with Neptune. It is pronounced "Q-B-one-O" and is named for the first such object discovered, minor planet (15760) 1992QB$_1$.

D/H ratio The ratio of deuterium to hydrogen in water, which can be used to determine the molecular source of the

	water – i.e., whether significant amounts of ocean water on Earth arrived from comets or other sources.
Deep Impact	The NASA space probe launched in 2005 that released an impactor that smashed into Comet 9P/Tempel 1. In 2007 it was redesignated *EPOXI* and flew past Comet 103P/Hartley 2.
Deep Space 1	A NASA spacecraft launched in 1998 that flew past Comet 19P/Borrelly in 2001.
"Dirty snowball"	The basic model of a cometary nucleus, proposed by astronomer Fred Whipple in 1950, laying out the structure of comets as blocks of ice infused with dust and particles of rock and dirt.
DNA	Deoxyribonucleic acid, a complex organic molecule that encodes and contains the genetic instructions used in the development and function of living organisms.
Dust tail	A long stream of dust particles pointing away from the Sun formed by particles of dust emerging from a comet's nucleus as its ices sublimate, visible as sunlight scatters off the dust.
Dwarf planet	A term adopted in 2006 by the International Astronomical Union to describe Pluto, Ceres, Haumea, Makemake, and Eris – solar system bodies in a solar orbit that are massive enough to have a gravitationally controlled shape but without having cleared their orbits of other celestial bodies.
Eccentricity	The amount by which a solar system body's orbit deviates from a circle.
Edgeworth-Kuiper Belt	An alternative name for simply Kuiper Belt (*see*), sometimes used because astronomer Kenneth Edgeworth was an astronomer who hypothesized such a region.
Elliptical orbit	An orbit with an eccentricity less than 1, which therefore plays out an ellipse.
Ephemeris	A list of predicted positions for a comet or other object moving against the background stars, typically generated by the International Astronomical Union's Minor Planet Center, through its Central Bureau for Astronomical Telegrams.

EPOXI — The renamed 2007-and-beyond phase of NASA's *Deep Impact* spacecraft, which witnessed a flyby of Comet 103P/Hartley 2.

Exocomet — A comet outside our solar system.

Gas giant planet — A large planet composed of gas rather than rock; in the solar system, Jupiter, Saturn, Uranus, and Neptune (although Uranus and Neptune are sometimes characterized as ice giants because they are composed mostly of materials less volatile than hydrogen and helium).

Gas tail — *See* ion tail.

Giant molecular cloud — An interstellar cloud composed of molecules – largely molecular hydrogen – that lead to star formation, the close passage of which near our solar system can gravitationally steer comets inward toward the Sun.

Giotto — A European spacecraft that explored Comet 1P/Halley, conducting a close flyby in 1986 after having been launched the previous year.

Glycine — The simplest of all amino acids, building blocks of proteins and important organic molecules, discovered in Comet 81P/Wild 2, in samples returned to Earth from the *Stardust* mission.

Great Comet — An exceptionally bright comet.

Halley-type comet — Short-period comet, like 1P/Halley, with period less than 200 years and orbital inclination extending from 0° to more than 90°.

Hill sphere — The gravitational region of a body in which other objects are attracted; named for astronomer George W. Hill.

Hipparcos — A European Space Agency satellite, launched in 1989, that operated until 1993, precisely measuring the positions of stars and numerous other celestial bodies.

Hyperbolic orbit — An orbit with eccentricity greater than 1; a comet on this trajectory will swoop in close to the Sun once, gliding off and never returning.

Inclination — The angle of a solar system body's orbit with respect to the ecliptic, the plane of the solar system's planets.

International Astronomical Union Abbreviated IAU, this international body of astronomers acts as the authority for naming and assigning designations to celestial bodies.

International Cometary Explorer Abbreviated *ICE*, this NASA and European satellite was launched in 1978 and flew through the tail of Comet 21P/Giacobini-Zinner in 1985 before studying Comet 1P/Halley in 1986.

International Halley Watch Organized by scientists at the Jet Propulsion Laboratory in California, this effort organized and compiled voluminous observations of Comet 1P/Halley in 1985 and 1986, creating a substantial archive.

Ion tail The gaseous molecules that sublimate away from a comet's nucleus are ionized by sunlight striking them, causing them to glow by fluorescence and by reflected sunlight. This forms the comet's primary tail and is often bluish in color. Also called a gas or plasma tail.

ISON The abbreviation for a Russian-led network of telescopes, the International Scientific Optical Network, that searches for comets and other small bodies in the solar system, its most famous discovery being Comet ISON (C/2012 S1).

Jupiter-family comet A short-period comet with a period less than 20 years and an inclination of 30° or less, gravitationally influenced by Jupiter.

Kreutz Sungrazer A member of a family of sungrazing comets characterized by orbits that carry them extremely close to the Sun at perihelion. Named for German astronomer Heinrich Kreutz.

Kuiper Airborne Observatory A NASA-funded observatory placed into a Lockheed C-141A Starlifter jet that was commissioned in 1974 and flew until 1995, making observations of numerous solar system bodies in the infrared part of the spectrum. Named for Dutch-American astronomer Gerard P. Kuiper.

Kuiper Belt A region of small solar system bodies beyond the planets, extending from the orbit of Neptune (at about 30 astronomical units from the Sun) to approximately 50 astronomical units from the Sun. Named for Dutch-American astronomer Gerard P. Kuiper.

Late-Heavy Bombardment A hypothesized period 4.1 billion to 3.8 billion years ago, during which Earth and other inner planets and moons underwent a large number of impacts from small bodies in the solar system.

LINEAR Acronym for Lincoln Near-Earth Asteroid Research, a program of discovery set up by NASA, the U.S. Air Force, and the Massachusetts Institute of Technology's Lincoln Laboratory, to discover and track asteroids systematically.

Long-period comet A comet with a period greater than 200 years, typically thousands or even millions, and with a highly eccentric orbit.

Main-belt comet Small solar system body orbiting within the main asteroid belt that has shown signs of cometary activity.

Meteor A bright streak of light caused by the ionization trail of a small particle – a meteoroid – entering Earth's atmosphere at a high velocity.

Meteor shower A concentration of meteors seemingly originating in a discrete area of sky, caused by Earth passing through a stream of particles left by the wake of a comet or asteroid.

Meteorite The physical fragment of an asteroid or comet that falls to Earth and is recovered; stony, iron, or a combination thereof.

Meteoroid An asteroidal or cometary particle in the solar system that may enter Earth's atmosphere, causing a bright meteor streak in our sky.

Miller-Urey experiment An experiment conducted in 1952 at the University of Chicago by Stanley Miller and Harold Urey that demonstrated the synthesis of organic compounds from inorganic precursors.

Minor Planet Center Under the auspices of the International Astronomical Union, the MPC operates at Smithsonian Astrophysical Observatory in Cambridge, Massachusetts, and is charged with the calculation of orbits and other data for comets and asteroids.

Near-Earth Object An NEO is a solar system body whose orbit carries it into close proximity to Earth.

NEAT — The acronym for Near-Earth Asteroid Tracking, a program run by NASA to discover and track near-Earth objects.

Nongravitational force — Force that acts on cometary nuclei aside from the Sun's gravitation, such as a jet created by warming gases and dust.

Nucleus — The frozen block of "dirty ice" that constitutes a comet, typically measuring a few kilometers across, and that warms up and sublimates ices and scatters dust when it moves into the inner solar system.

Oort Cloud — A hypothesized shell of comets surrounding the solar system that may be centered on space some 50,000 or more astronomical units, or nearly a light-year, away from the Sun – a quarter of the way to the nearest star. Named for Dutch astronomer Jan H. Oort.

Panspermia — A hypothesis that microbial life exists throughout the universe and is carried by comets, asteroids, meteoroids, and planetoids.

PANSTARRS — The acronym for Panoramic Survey Telescope and Rapid Response System (Pan-STARRS), a planned array of telescope facilities, the first of which operates on the Hawaiian island of Maui. Cometary designations by the IAU do not use the telescope name, Pan-STARRS, but rather PANSTARRS.

Parabolic orbit — The orbit of a celestial body with the eccentricity equal to 1, forming a parabola around, say, the Sun.

Perigee — The closest point a solar system body reaches to Earth in its orbit.

Perihelion — The nearest point to the Sun in a solar system object's orbit.

Periodic comet — A comet that returns to a point close to the Sun in a measurable period.

Perturbation — Change in the orbit of a solar system object caused by the gravity of objects other than the Sun – and typically by Jupiter.

Planetesimal — A protoplanet that is forming and growing larger by accretion in an early stage of its development. They are

	thought to have existed in large numbers in the early solar system and in other forming solar systems.
Plasma	Matter consisting of positively charged ions and negatively charged electrons.
Plasma tail	*See* ion tail.
Plutino	A Trans-Neptunian Object in a 2:3 orbital resonance with Neptune.
Polycyclic aromatic hydrocarbon	A PAH is a large, complex organic compound that occurs in comets, in the interstellar medium, and in oil, tar, and coal deposits on Earth.
Poynting-Robertson effect	An effect wherein solar radiation causes dust particles in the solar system to be slowly spiraled into the Sun.
Protein	A large biological molecule consisting of one or more chains of amino acids.
Protosolar disk	The rotating disk of matter thought to exist in the early solar system that led to the accretion of the Sun, planets, and small bodies such as comets.
Protosolar nebula	The nebular cloud that formed by gravity in the earliest days of our solar system leading to a rotating disk, the Sun, planets, and small bodies such as comets.
Radiation pressure	Electromagnetic radiation from the Sun pushing particles outward, away from the center of the solar system.
Refractories	Dusty particles in comets that have relatively high condensation temperatures, including metal and silicate grains.
Resonance	The effect of two bodies exerting a regular, periodic, gravitational influence on each other, such that their orbital periods can be related in a frequency by two integers.
RNA	Ribonucleic acid, a family of large biological molecules that perform vital roles in the coding of genetic material.
Rosetta	A spacecraft launched by the European Space Agency in 2004 designed to study and land on Comet 67P/Churyumov-Gerasimenko in 2014.
Sakigake	A Japanese space probe that studied Comet 1P/Halley in 1986.

Scattered disk A distant region of the outer solar system that is sparsely populated by icy bodies – a subset of Trans-Neptunian Objects.

Semimajor axis One-half of the long axis of the ellipse of an orbit.

Short-period comet Periodic comet with a period of less than 200 years.

SOHO The acronym for the joint ESA/NASA *Solar and Heliospheric Observatory*, launched in 1995 and responsible for imaging many close encounters between comets and the Sun.

Solar wind A stream of charged particles emitted from the Sun's upper atmosphere.

Stardust A NASA spacecraft launched in 1999 that encountered and sampled Comet 81P/Wild 2 in 2004, returning samples for study in 2006. Renamed *Stardust-NExT*, the craft then encountered Comet 9P/Tempel 1 in 2011.

Sublimation The act in which matter changes from a solid into a gas without going through a liquid stage.

Suisei A Japanese space probe that encountered Comet 1P/Halley in 1986.

Sungrazer A comet that passes extremely close to the Sun at perihelion.

Trans-Neptunian Object A TNO is an object in the solar system that orbits the Sun at a greater semimajor axis than that of Neptune and includes many icy bodies such as comets.

Tunguska Event An extremely powerful explosion that occurred June 30, 1908, over the Podkamennaya Tunguska River in Siberia, thought to have been caused by the airburst of a cometary nucleus.

Twotino A Trans-Neptunian Object with a 1:2 orbital resonance with Neptune.

Vega 1 and Vega 2 Soviet space probes that studied Comet 1P/Halley in 1986.

Volatiles Icy materials in comets that have low boiling points and sublime away to form comae and tails. They include elements and compounds such as water, ammonia, methane, carbon dioxide, nitrogen, and hydrogen.

VSMOW The acronym for Vienna Standard Mean Ocean Water, a standard defining the isotopic composition of Earth's ocean water.

Bibliography

Andrews, Bill, ed. *Explore the Solar System*. Second ed. Waukesha, Wis.: Kalmbach, 2012.

Bailey, M. E., S. V. M. Clube, and W. M. Napier, *The Origin of Comets*. New York: Pergamon Press, 1990.

Boehnhardt, H., M. Combi, M. R. Kidger, and R. Schulz, eds. *Cometary Science after Hale-Bopp*. Two vols. Boston: Kluwer Academic, 2002.

Brandt, John C., and Robert D. Chapman. *Rendezvous in Space: The Science of Comets*. New York: W. H. Freeman, 1992.

Burnham, Robert. *Comet Hale-Bopp: Find and Enjoy the Great Comet*. New York: Cambridge University Press, 1997.

Burnham, R. *Great Comets*. New York: Cambridge University Press, 2000.

Beatty, J. Kelly, Carolyn Collins Petersen, and Andrew Chaikin, eds. *The New Solar System*. Fourth ed. Cambridge, Mass.: Sky Publishing Corp. and Cambridge University Press, 1999.

Brandt, John C., and Robert D. Chapman. *Introduction to Comets*. Second ed. New York: Cambridge University Press, 2004.

Crovisier, Jacques, and Thérèse Encrenaz, *Comet Science: The Study of Remnants from the Birth of the Solar System*. New York: Cambridge University Press, 2000.

Deskins, David. *Looking Back: Amateur Adventures with Halley's Comet, 1985–1986*. Pikeville, Ky.: Intrinsic, 1987.

Dickinson, Terence, and Alan Dyer. *The Backyard Astronomer's Guide*. Third ed. Buffalo, New York: Firefly Books, 2010.

Edberg, Stephen J., and David H. Levy. *Observing Comets, Asteroids, Meteors, and the Zodiacal Light*. New York: Cambridge University Press, 1994.

Festou, M. C., H. U. Keller, and H. A. Weaver, eds. *Comets II*. Tucson: University of Arizona Press, 2004.

Gehrels, Tom, ed. *Hazards Due to Comets and Asteroids*. Tucson: University of Arizona Press, 1994.

Grewing, Michael, Françoise Praderie, and Rüdeger Reinhard, eds. *Exploration of Halley's Comet*. New York: Springer-Verlag, 1988.

Hale, Alan. *Everybody's Comet: A Layman's Guide to Hale-Bopp*. Silver City, N.M.: High Lonesome Books, 1996.

Huebner, Walter F., ed. *Physics and Chemistry of Comets*. New York: Springer-Verlag, 1990.

James, Nick, and Gerald North. *Observing Comets*. New York: Springer-Verlag, 2003.

Kronk, Gary W. *Cometography*. Five vols. New York: Cambridge University Press, 1999–2010.

Kronk, G. W. *Comets: A Descriptive Catalog*. Hillside, N.J.: Enslow, 1984.

Levy, David H. *Comets: Creators and Destroyers*. New York: Touchstone, 1998.

Levy, D. H. *David Levy's Guide to Observing and Discovering Comets*. New York: Cambridge University Press, 2003.

Impact Jupiter: The Crash of Comet Shoemaker-Levy 9. New York: Plenum Press, 1995.

Lewis, John S. *Rain of Iron and Ice: The Very Real Threat of Comet and Asteroid Bombardment*. Reading, Mass.: Addison-Wesley, 1996.

Littmann, Mark, and Donald K. Yeomans. *Comet Halley: Once in a Lifetime*. Washington, D.C.: American Chemical Society, 1985.

McFadden, Lucy Ann, Paul Weissman, and Torrence Johnson, eds. *Encyclopedia of the Solar System*. Second ed. Waltham, Mass.: Academic Press, 2006.

Mobberley, Martin. *Hunting and Imaging Comets*. New York: Springer-Verlag, 2011.

Moore, Patrick, and John Mason. *The Return of Halley's Comet*. New York: Warner Books, 1985.

Rahe, Jürgen, Bertram Donn, and Karl Wurm. *Atlas of Cometary Forms: Structures near the Nucleus*. Washington, D.C.: U.S. Government Printing Office, 1969.

Reddy, Francis. *Halley's Comet!* Milwaukee: AstroMedia, 1985.

Schechner, Sara. *Comets, Popular Culture, and the Birth of Modern Cosmology*. Princeton, N.J.: Princeton University Press, 1999.

Schmude, Richard. *Comets and How to Observe Them*. New York: Springer-Verlag, 2010.

Seargent, David A. J. *The Greatest Comets in History: Broom Stars and Celestial Scimitars*. New York: Springer-Verlag, 2008.

Spencer, John R., and Jacqueline Mitton, eds. *The Great Comet Crash: The Impact of Comet Shoemaker-Levy 9 on Jupiter*. New York: Cambridge University Press, 1995.

Yeomans, Donald K. *Comets: A Chronological History of Observation, Science, Myth, and Folklore*. New York: John Wiley and Sons, 1991.

Index